AGN FEEDBACK IN GALAXY FORMATION

During the past decade, convincing evidence has been accumulated concerning the effect of active galactic nuclei (AGN) activity on the internal and external environment of their host galaxies. Featuring contributions from well-respected researchers in the field, and bringing together work by specialists in both galaxy formation and AGN, this volume addresses a number of key questions about AGN feedback in the context of galaxy formation.

The topics covered include downsizing and star-formation timescales in massive elliptical galaxies, the connection between the epochs of supermassive black hole growth and galaxy formation and the question of whether AGN and star formation coexist. The book also discusses key challenging computational problems, including jet–interstellar/intergalactic medium interactions, and both jet- and merging-induced star formation.

Suitable for both researchers and graduate students in astrophysics, this volume reflects the engaging and lively discussions taking place in this emerging field of research.

VINCENZO ANTONUCCIO-DELOGU is a research astronomer at the Istituto Nazionale di Astrofisica, Osservatorio Astrofisico di Catania, Italy. His research focuses on dynamics and substructure in clusters of galaxies, parallel N-body simulation codes and galaxy formation and evolution.

JOSEPH SILK is Savilian Professor of Astronomy at the University of Oxford, and Director of the Beecroft Institute of Particle Astrophysics and Cosmology (BIPAC). His research interests include theoretical cosmology, dark matter, galaxy formation and the cosmic microwave background.

AGN FEEDBACK
IN GALAXY FORMATION

Proceedings of the Workshop held in Vulcano,
Italy, May 18–22, 2008

Edited by

VINCENZO ANTONUCCIO-DELOGU
Istituto Nazionale di Astrofisica

JOSEPH SILK
University of Oxford

CAMBRIDGE
UNIVERSITY PRESS

CAMBRIDGE
UNIVERSITY PRESS

University Printing House, Cambridge CB2 8BS, United Kingdom

One Liberty Plaza, 20th Floor, New York, NY 10006, USA

477 Williamstown Road, Port Melbourne, VIC 3207, Australia

314-321, 3rd Floor, Plot 3, Splendor Forum, Jasola District Centre, New Delhi - 110025, India

103 Penang Road, #05-06/07, Visioncrest Commercial, Singapore 238467

Cambridge University Press is part of the University of Cambridge.

It furthers the University's mission by disseminating knowledge in the pursuit of education, learning and research at the highest international levels of excellence.

www.cambridge.org
Information on this title: www.cambridge.org/9780521192545

First published 2011

A catalogue record for this publication is available from the British Library

Library of Congress Cataloging in Publication data
AGN feedback in galaxy formation : proceedings of the workshop held in Vulcano, Italy,
May 18–22, 2008 / edited by V. Antonuccio-Delogu and J. Silk.
p.   cm. – (Cambridge contemporary astrophysics)
ISBN 978-0-521-19254-5 (hardback)
1. Galaxies – Evolution – Mathematical models – Congresses.   2. Galaxies – Formation –
Mathematical models – Congresses.   3. Galactic nuclei – Congresses.   I. Antonuccio-Delogu,
V. (Vincenzo), 1960–   II. Silk, Joseph, 1942–   III. Title.   IV. Series.
QB857.5.E96   A35
523.1′12 – dc22   2010024385

ISBN 978-0-521-19254-5 Hardback

Contents

List of contributors   *page* x
Preface   xiii
The organising committees   xv

Part I   AGNs, starbursts and galaxy evolution   **1**
1   The effects of mass and star-formation timescale on
     galaxy evolution   3
     C. D. Harrison and M. Colless
     1.1   Introduction   3
     1.2   Estimating the stellar population parameters   4
     1.3   Scaling relations   4
     1.4   Parameter distributions   6
     1.5   The influence of mass   7
     1.6   Star formation inside out   8
     1.7   Summary   10
            References   10
2   Suppressing cluster cooling flows by multiple AGN activity   11
     A. Nusser
     2.1   Introduction   11
     2.2   Outline of the model   12
     2.3   Results   17
     2.4   Summary and discussion   18
            References   20
3   Starburst and AGN activity in Spitzer-selected sources at high-z 21
     M. Polletta, A. Omont, C. Lonsdale and D. Shupe
     3.1   Introduction   21
     3.2   Spitzer selection of high-z luminous infrared galaxies   21

3.3   MAMBO observations and results   23
3.4   AGN and starburst MIR emission contributions   24
3.5   Host galaxies   25
3.6   Average SEDs   26
3.7   Summary and conclusions   27
      References   28

4   Star formation in galaxies hosting active galactic nuclei up to $z \sim 1$ 29
     J. D. Silverman, zCOSMOS and XMM-COSMOS
4.1   Introduction   29
4.2   Star formation rates in zCOSMOS galaxies hosting AGN   32
4.3   Further remarks on color–magnitude diagrams of AGN hosts   33
4.4   Conclusion: co-evolution of SMBHs and their host galaxies   35
      References   36

Part II   Co-evolution of black holes and galaxies   **39**
5   The symbiosis between galaxies and SMBHs   41
     G. L. Granato, M. Cook, A. Lapi and L. Silva
5.1   Introduction   41
5.2   Standard SAMs, their successes and their failures   41
5.3   Possible solution from joint evolution of QSO and spheroids   43
5.4   The ABC scenario   44
      References   45

6   On the origin of halo assembly bias   47
     A. Keselman
6.1   Introduction   47
6.2   Measuring assembly bias in the quasi-linear regime   49
6.3   Conclusions   50
      References   50

7   AGN, downsizing and galaxy bimodality   52
     M. J. Stringer, A. J. Benson, K. Bundy and R. S. Ellis
7.1   Introduction   52
7.2   Hierarchical assembly   52
7.3   Modelling AGN feedback in galaxies   54
7.4   Colour bimodality   56
7.5   Understanding mass errors and cosmic variance   56
      Acknowledgements   59
      References   59

Part III   Outflows and radio galaxies   **61**
8   Interaction and gas outflows in radio-loud AGN: disruptive and
     constructive effects of radio jets   63
     R. Morganti
8.1   Why radio-loud AGN?   63

Contents   vii

8.2	The nuclear regions probed by the HI and ionised gas	65
8.3	Moving to larger scales: jet-induced star formation	69
8.4	Conclusions	72
	Acknowledgements	73
	References	73

9 Young radio sources: evolution and broad-band emission   75
   L. Ostorero, R. Moderski, Ł. Stawarz, M. Begelman, A. Diaferio,
   I. Kowalska, J. Kataoka and S. J. Wagner

9.1	Introduction	75
9.2	The model: dynamical and spectral evolution	75
9.3	Comparison with broad-band spectra of GPS galaxies	77
9.4	Further observational support	78
9.5	Conclusions and future prospects	80
	Acknowledgements	81
	References	81

10 The duty cycle of radio galaxies and AGN feedback   82
    S. Shabala

10.1	Introduction	82
10.2	Local sample	83
10.3	Intermittent AGN feedback in galaxy formation	88
10.4	Summary	95
	Acknowledgements	96
	References	96

11 Environment or outflows? New insight into the origin of narrow
    associated QSO absorbers   98
    V. Wild

11.1	Introduction	98
11.2	Using ultraviolet NALs to reveal QSO feedback	99
11.3	The line-of-sight distribution of NALs in front of QSOs	101
11.4	The 3D distribution of NALs around QSOs	102
11.5	The clustering contribution to the line-of-sight excess	104
11.6	Radio loud vs. radio quiet	105
11.7	Conclusions	106
	Acknowledgements	107
	References	107

**Part IV   Models and numerical simulations: methods
             and results**   **109**

12 Physical models of AGN feedback   111
    V. Antonuccio-Delogu, J. Silk, C. Tortora, S. Kaviraj N. Napolitano
    and A. D. Romeo

12.1	Introduction	111

Contents

12.2	Simulating jet propagation in a two-phase ISM	112
12.3	Global quenching in elliptical galaxies	129
12.4	Conclusions	151
	References	154
13	Large-scale expansion of AGN outflows in a cosmological volume	157
	P. Barai	
13.1	Introduction	157
13.2	The numerical setup	157
13.3	Results and discussion	160
	References	163
14	Relativistic jets and the inhomogeneous interstellar medium	165
	G. V. Bicknell, J. L. Cooper and R. S. Sutherland	
14.1	AGN feedback from a radio galaxy perspective	165
14.2	Simulation code	166
14.3	Isotropisation of jet momentum	166
14.4	Jet and disk simulations	166
14.5	Application to 4C31.04	169
14.6	Interaction of outflows with individual clouds	171
14.7	Main points	172
	References	174
15	AGN feedback effect on intracluster medium properties from galaxy cluster hydrodynamical simulations	175
	D. Fabjan, S. Borgani, L. Tornatore, A. Saro and K. Dolag	
15.1	Introduction	175
15.2	The simulations	176
15.3	Temperature profiles	177
15.4	Metal enrichment of the ICM	178
15.5	Conclusions	180
	Acknowledgements	181
	References	182
16	Physics and fate of jet-related emission line regions	183
	M. G. H. Krause and V. Gaibler	
16.1	Introduction	183
16.2	Global jet simulations	185
16.3	Local simulations of multi-phase turbulence	187
16.4	Discussion and conclusions	190
	References	192
17	Cusp–core dichotomy of elliptical galaxies: the role of thermal evaporation	194
	C. Nipoti	

Contents   ix

17.1 Introduction   194
17.2 The formation of cusps and cores in elliptical galaxies   195
17.3 Implications for active galactic nuclei in elliptical galaxies   197
17.4 Conclusions   198
     Acknowledgements   198
     References   198

Index   200

Contributors

V. Antonuccio-Delogu, Istituto Nazionale di Astrofisica (INAF) Osservatorio Astrofisica di Catania, Via S. Sofia 78, 95123 Catania, Italy

P. Barai, University of Nevada, Las Vegas, Department of Physics & Astronomy, 4505 S. Maryland Parkway, Box 454002, Las Vegas, NV 89154-4002, USA

M. Begelman, JILA, University of Colorado, Boulder, CO 80309-0440, USA

A. J. Benson, Caltech, 1200 E. California Blvd., Pasadena, CA 91125, USA

G. V. Bicknell, Australian National University, Research School of Astronomy & Astrophysics, Mt Stromlo Observatory, Cotter Rd, Weston ACT 2611, Australia

S. Borgani, University of Trieste, Department of Astronomy, Via Tiepolo 11, 34134 Trieste, Italy

K. Bundy, University of California, Berkeley, Astronomy Department, 601 Campbell Hall, Berkeley, CA 94720-3411, USA

M. Colless, Anglo-Australian Observatory, PO Box 296, Epping, NSW 2121, Australia

M. Cook, SISSA-ISAS, Trieste, Via Beirut 2-4, 34151 Trieste, Italy

J. L. Cooper, Australian National University, Research School of Astronomy & Astrophysics, Mt Stromlo Observatory, Cotter Rd, Weston ACT 2611, Australia,

A. Diaferio, Università degli Studi di Torino, Dipartimento di Fisica, Via Giuria 1, 10125, Torino, Italy

K. Dolag, Max Planck Institute für Astrophysik, Karl-Schwarzschild Strasse 1, Garching bei Muenchen, Germany

R. S. Ellis, University of Oxford, Oxford Astrophysics, Keble Road, Oxford OX1 3RH, UK

D. Fabjan, Osservatorio Astronomico di Trieste, Via Tiepolo 11, 34143 Trieste, Italy

V. Gaibler, Max Planck Institute für Extraterrestrische Physik, Postfach 1312, 85741 Garching, Germany

G. L. Granato, INAF, Osservatorio Astronomico di Trieste, Via Tiepolo 11, 34143 Trieste, Italy

C. D. Harrison, Cerro Tololo Inter-American Observatory, Casilla 603, La Serena, Chile

J. Kataoka, Tokyo Institute of Technology, Department of Physics, 2-12-1, Ohokayama, Meguro Tokyo, 152-82551, Japan

S. Kaviraj, University of Oxford, Oxford Astrophysics, Denys Wilkinson Building, Keble Road, Oxford, OX1 3RH, UK

A. Keselman, Technion, Physics Department, Haifa 32000, Israel

I. Kowalska, University of Warsaw, Astronomical Observatory, Al. Ujazdowskie 4, 00–478 Warsaw, Poland

M. G. H. Krause, Max Planck Institute für Extraterrestrische Physik, Postfach 1312, 85741 Garching, Germany

A. Lapi, University of Rome 'Tor Vergata', Department of Physics, Via Della Ricerca Scientifica 1, 00133 Roma, Italy

C. Lonsdale, University of Virginia, Charlottesville, VA 22904, USA

R. Moderski, Nicolaus Copernicus Astronomical Centre, Bartycka 18, 00-716 Warsaw, Poland

R. Morganti, ASTRON, PO Box 2, 7990 AA Dwingeloo, The Netherlands

N. Napolitano, INAF, Osservatorio Astronomico di Capodimonte, Salita Moiariello 16, Naples, Italy

C. Nipoti, University of Bologna, Astronomy Department, Via Ranzani, 1, 40127 Bologna, Italy

A. Nusser, Technion, Physics Department, Haifa 32000, Israel

A. Omont, IAP, 75014 Paris, France

L. Ostorero, Università degli Studi di Torino, Dipartimento di Fisica, Via Giuria 1, 10125 Torino, Italy

M. Polletta, INAF-IASF Milano, Via E. Bassini 15, 20133 Milano, Italy

A. D. Romeo, Universidad Andres Bello, Departamento Ciencias Fisicas, Republica 252 Santiago, Chile

A. Saro, University of Trieste, Department of Astronomy, Via Tiepolo 11, 34134 Trieste, Italy

S. Shabala, University of Oxford, Oxford Astrophysics, Denys Wilkinson Building, Keble Road, Oxford, OX1 3RH, UK

D. Shupe, IPAC, Pasadena, CA 91125, USA

J. Silk, University of Oxford, Oxford Astrophysics, Denys Wilkinson Building, Keble Road, Oxford, OX1 3RH, UK

L. Silva, Osservatorio Astronomico di Trieste, Via Tiepolo 11, 34143 Trieste, Italy

J. D. Silverman, ETH, Institute of Astronomy, HIT J13.2, Wolfgang-Pauli-Strasse 27, 8093, Zurich, Switzerland

Ł. Stawarz, Stanford University, Kavli Research Institute, Stanford, CA 94305, USA

M. J. Stringer, Durham University, Department of Physics, South Road, Durham, DH1 3LE, UK

R. S. Sutherland, Australian National University, Research School of Astronomy & Astrophysics, Mt Stromlo Observatory, Cotter Rd, Weston ACT 2611, Australia

L. Tornatore, University of Trieste, Department of Astronomy, Via Tiepolo 11, 34134 Trieste, Italy

C. Tortora, INAF, Osservatorio Astronomico di Capodimonte, Salita Moiariello 16, Naples, Italy

S. J. Wagner, Landessternwarte, 69117 Heidelberg, Germany

V. Wild, Institut d'Astrophysique de Paris, 98bis Boulevard Arago, 75014 Paris, France

Preface

During the past decade, convincing evidence has been accumulated concerning the effect that AGN activity has on the internal and external environment of host galaxies. At intermediate and relatively high redshifts (z-0.2–1.5) evidence for this interaction comes, for example, from the optical–radio alignment and from the observation of jet-induced star formation. In the nearby universe there is also a series of significant indications: the observation of recent episodes of star formation in otherwise old or early types of ellipticals has emerged from analyses of the SDSS. There is also more direct and circumstantial evidence from the analysis of regions such as the Minkowski object, or the distribution of star-forming regions around the nearby radio envelope of Cen A, and from the enhanced star formation seen in some satellite galaxies of active galaxies at relatively high redshift.

Parallel and somewhat independently from this more direct evidence, the study of galaxy evolution has provided the astrophysical community with challenging new questions. The availability of large-scale photometric and spectral surveys such as the 2dF and the Sloan Digital Sky Survey has made it possible to discover evidence for evolution of the stellar formation features on timescales that are very short, in cosmological terms. The paradigm thus emerging in the astrophysical community is that AGN activity could be tightly connected to these phenomena, and could be capable of affecting the evolution of stellar populations within galaxies.

The purpose of the Oxford–COSMOCT workshop on *The Interface Between Galaxy Formation and AGNs*, which took place on the island of Vulcano, Italy, from May 18th to 22nd, 2008, was to bring together two communities, studying galaxy formation and AGNs, with a view to better understanding AGN feedback in the context of galaxy formation. The observational connection also included more specific observational and theoretical evidence, such as jet-induced star formation, and the association of starbursts with AGNs and superwinds. The Scientific Committee put special emphasis on some central questions, which included the following: Is AGN feedback necessary to appreciate why the most massive galaxies are red and dead? How do we understand downsizing and star-formation timescales in massive ellipticals? Can AGN provide positive as well as negative feedback for galaxy formation? What is the connection between the epochs of SMBH growth and galaxy formation? What is the evidence for jet-induced star formation? Do AGNs and star formation coexist, and is there a causal connection? This volume collects the proceedings presented by most participants, and reflects the lively discussions on the observational and computational problems connected to the phenomenology of AGN feedback on their host galaxies. Particular care has been taken in discussing some key challenging computational problems, including (among others) jet–interstellar/intergalactic medium interactions, jet-induced stellar formation, and merging-induced stellar formation.

The subject of AGN feedback on their host galaxies, with all its rich observational phenomenology, is a relatively young one. Many different phenomena, such as the massive outflows from post-starburst galaxies, galaxy colour bimodality and others, can be projected into the perspective of the mutual interaction between AGN activity and galaxy formation. The workshop was the first one specifically dedicated to discussing this emerging paradigm, and we believe that the efforts made by the participants and by those who contributed to this volume will be useful for the astrophysical community at large.

The generous contributions from our sponsors, mentioned below, provided the necessary resources to organise this event. Special acknowledgment should be given to the technical and administrative staff of INAF, Catania Astrophysical Observatory, and particularly to Luigia Santagati for her dedication during the preparatory phases, and to Alfio Giuffrida and Piero Massimino, who brought a decent WiFi IT connection to the workshop's site, and cared especially that it would work efficiently for the entire duration. We would also like to thank the secretarial staff, and in particular Stavro Ivanovski and Alessio Romeo, who kindly provided their help always with a smile.

The organising committees

Scientific Organising Committee
J. Silk (Oxford, UK, *Chair*)
F. Combes (Paris, France)
G. Hasinger (Heidelberg, Germany)
C. Norman (STSCI, USA)
V. Antonuccio-Delogu (Catania, Italy)

Local Organising Committee
V. Antonuccio-Delogu
Alessio D. Romeo
Alfio Giuffrida
Stavro Ivanovski
Luigia Santagati

Acknowledgements   The workshop was made possible by generous donations from the European Science Foundation programme ASTROSIM – European Network for Computational Astrophysics (www.astrosim.net/), from the European Commission VI Framework Programme for Research & Development, *Transfer of Knowledge* Project (contract MTKD-CT-002995, *Cosmology and Computational Astrophysics at Catania Astrophysical Observatory*), and from the Italian National Institute for Astrophysics, INAF.

Part I

AGNs, starbursts and galaxy evolution

1

The effects of mass and star-formation timescale on galaxy evolution

Craig D. Harrison & Matthew Colless

1.1 Introduction

In a ΛCDM cosmology, the growth of structure occurs hierarchically; small objects form first and undergo mergers to form more massive objects. Modelling the formation and evolution of galaxies with numerical simulations is impossible because crucial aspects lack a complete physical model – notably, feedback and star formation.

An alternative to full numerical simulations is the semi-analytic models (Croton *et al.* 2006; De Lucia *et al.* 2006), so called because they use approximate prescriptions for physical processes that are poorly understood. These prescriptions contain parameters that are set by demanding that the model reproduces the observations of (typically) low-redshift galaxies. The process itself is often motivated by a result from a more detailed numerical simulation or from observations.

The cooling of gas is central to the process of galaxy formation, as it sets the rate at which gas becomes available for star formation. Feedback processes have the largest impact on the predictions for galaxy properties, as these processes affect the efficiency of galaxy formation by increasing the cooling time of hot gas and suppressing further star formation. Previous iterations of the semi-analytic models (Kauffmann 1996), which lacked a prescription for feedback, predicted that the galaxy population should continue to evolve at $z < 1$, at a rate greater than that due to passive evolution, as more massive galaxies are built up by continued mergers. These mergers of galaxies with gas reserves (wet mergers) are accompanied by a burst of star formation and, consequently, massive galaxies are predicted to exhibit younger ages and higher metallicities than less massive galaxies. More recent models incorporating 'radio mode' feedback from AGN, which suppresses star formation in massive galaxies at higher redshifts (Croton *et al.* 2006), predict that while early-type galaxies assemble their mass hierarchically they have

AGN Feedback in Galaxy Formation, eds. V. Antonuccio-Delogu and J. Silk. Published by Cambridge University Press. © Cambridge University Press 2011.

anti-hierarchical star-formation histories (De Lucia *et al.* 2006). In these models, massive early-type galaxies are predicted to be older, more metal-rich, and have higher α-element abundance ratios than less massive galaxies, even though their mass assembly was completed at a later time through purely stellar mergers (dry mergers). Trends of decreasing mass, age and metallicity with increasing cluster-centric distance are also predicted.

We test the predictions from these more recent models by analysing the stellar populations of cluster early-type galaxies. Redshifts, velocity dispersions and Lick absorption-line strengths (Burstein *et al.* 1984; Trager *et al.* 1998) are measured for ~100 galaxies from four clusters. The ages, metallicities and α-element abundance ratios of these galaxies are estimated by comparison of the absorption-line strengths with up-to-date stellar population models.

1.2 Estimating the stellar population parameters

The stellar population parameters, age (t), metallicity ($[Z/H]$) and α-element abundance ratio ($[\alpha/Fe]$), were estimated for ~100 galaxies in four low-z clusters. These clusters have $\langle z \rangle = 0.04$ and were selected to span the range of both Abell richness classes and B-M classifications.

Along with the redshift (z) and velocity dispersion (σ), the strengths of three Lick absorption-line indices, Hβ, Mg b, and Fe5335, were measured for each galaxy. These strengths were then corrected for σ broadening before being calibrated to the Lick system.

The stellar population parameters were estimated using a method similar to Proctor *et al.* (2004) and compared to the predictions from the Thomas *et al.* (2003) single stellar population models. The models were interpolated in steps of 0.25 Gyr in t and 0.025 dex in $[Z/H]$ and $[\alpha/Fe]$.

The ratio of α-elements to Fe is commonly interpreted as a measure of star-formation timescale. Fe is predominantly created in SNIa while α-elements are predominantly created in SNII. Since there is a delay between the onset of SNII and SNIa, $[\alpha/Fe]$ can be used as a chronometer to estimate the duration of star formation.

1.3 Scaling relations

Figure 1.1 shows the variations of t (left), $[Z/H]$ (middle), and $[\alpha/Fe]$ (right) with σ. Mean error bars are shown in the bottom right corner of each panel. The solid lines in each panel are the linear fits to the data (the dashed lines in the left panel are visual aids that will be discussed later). A significant correlation with σ is found for both $[Z/H]$ and $[\alpha/Fe]$. Performing linear fits to the relations, taking into account

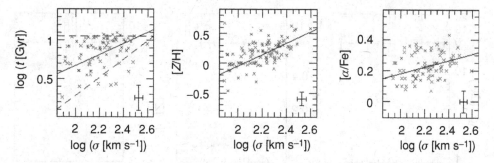

Figure 1.1 The variation of the stellar population parameters with σ. The solid line in each panel is the fit to the data. The dashed lines are simply guides for the eyes.

the errors in both quantities, we find

$$\log(t) = (0.72 \pm 0.09) \log \sigma - (0.77 \pm 0.21),$$

$$[Z/H] = (0.97 \pm 0.08) \log \sigma - (1.99 \pm 0.17),$$

$$[\alpha/Fe] = (0.21 \pm 0.05) \log \sigma - (0.24 \pm 0.12).$$

Looking at $[Z/H]$ first, we find that it is strongly correlated with σ ($r_S = 0.56$), at a significance level $>5\sigma$. A moderate correlation is found between $[\alpha/Fe]$ and σ ($r_S = 0.32$), a result that is significant at the 3σ level.

Given the size of the mean errors, the tightness of these two relations is remarkable, and implies that σ is a major factor in determining the $[Z/H]$ and $[\alpha/Fe]$ of a galaxy, with more massive galaxies being older and more metal-rich than less massive galaxies.

A moderate correlation is detected between t and σ ($r_S = 0.31$), significant at only the 2.8σ level. The distribution of points, however, is rather interesting. At all σ there exist galaxies with very old ages (marked by the horizontal dashed line), but the age of the youngest galaxy (at a given σ) increases with σ (approximately as the sloped, dashed line). This behaviour is reminiscent of down-sizing (Cowie *et al.* 1996), where the typical mass of a star-forming galaxy increases with z.

The possibility that a real correlation between t and σ exists cannot be discounted. The age limit of the models causes the build-up of galaxies on the upper edge of the distribution that may blur out any correlation and result in the conclusion of down-sizing. If this were the case, there would be an increase in the density of points on the upper edge of the distribution with increasing σ, which we do not see.

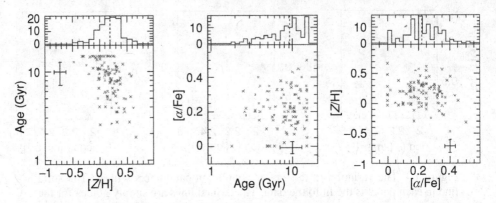

Figure 1.2 The stellar population parameter distributions and mean errors. Marginal distributions are given at the top of each panel with the median value marked with dotted lines.

Calculating ranges of the stellar population parameters predicted by these fits for a realistic range of σ, we find that they are in good agreement with the range of values observed for early-type galaxies in clusters (Trager *et al.* 2000).

These results confirm the predictions from De Lucia *et al.* (2006) that $[Z/H]$ and $[\alpha/Fe]$ are both positively correlated with mass. While we cannot confirm a correlation between t and mass, we do find evidence for down-sizing.

1.4 Parameter distributions

The distributions of stellar population parameters are shown in Figure 1.2. Marginal distributions are plotted for each of the parameters and the median values are marked as dotted lines. Median errors are given in each panel.

Looking at the distributions, we note that very few galaxies have $[Z/H]$ less than solar with most having $0.0 \lesssim [Z/H] \lesssim 0.5$ dex. Similarly, $[\alpha/Fe]$ is mostly greater than solar and lies in the range $0.1 \lesssim [\alpha/Fe] \lesssim 0.4$ dex. The bulk of the galaxies have old ages and almost all lie in the range $4 \lesssim t \lesssim 15$ Gyr. These distributions are in general agreement with those found by previous authors (Thomas *et al.* 2005; Collobert *et al.* 2006). The median $[Z/H]$ of our sample is 0.19 ± 0.02 dex, the median t is 9.2 ± 0.5 Gyr, and the median $[\alpha/Fe]$ is 0.22 ± 0.01 dex. The errors on these median values were determined from Monte Carlo simulations.

Looking at the marginal distributions, the $[Z/H]$ and $[\alpha/Fe]$ distributions are approximately Gaussian in nature. However, the t distribution is approximately exponential; most galaxies are old ($t \geq 8$ Gyr) but there is a tail of galaxies extending to ages as young as $\sim$2 Gyr.

It is also evident that there appears to be an anti-correlation between [Z/H] and t, in the sense that younger galaxies are more metal-rich. This might in principle be due to the non-orthogonal nature of the metallicity–age grids produced by the stellar population models, which means that the errors on these two quantities are correlated.

1.5 The influence of mass

The spread in the stellar population parameter distributions found for the low-z cluster sample is partly due to a combination of observational errors and the correlations with σ. To determine the amount of intrinsic scatter in the distributions, over and above that caused by the observational errors, we ran a series of numerical simulations. The method used was as follows.

Recalling that the t distribution is approximately exponential and that both the [Z/H] and [α/Fe] distributions are approximately Gaussian, we began by selecting a range of e-foldings for the exponential t distribution (τ) and a range of Gaussian scatters in [Z/H] and [α/Fe] ($\sigma_{[Z/H]}$ and $\sigma_{[\alpha/Fe]}$). The medians of the [Z/H] and [α/Fe] distributions were used as the means of their Gaussian distributions, and the exponential t distribution was truncated at 5 Gyr (at younger ages the models are less reliable) and at 14 Gyr (the age of the universe).

A model was generated for each combination of τ, $\sigma_{[Z/H]}$, and $\sigma_{[\alpha/Fe]}$ consisting of 10 000 mock galaxies with t, [Z/H] and [α/Fe] (parameter triples) drawn randomly from the specified distributions.

These parameter triples were converted to the corresponding Hβ, Mg b and Fe5335 index values (index triples). These index triples were then perturbed by randomly drawing errors from the observed galaxies' index error distributions. The perturbed index triples were converted back to parameter triples by exactly the same method used to estimate the stellar population parameters for the observed galaxies.

To ascertain how well the models fit the data a likelihood statistic was used. The stellar population parameter space was divided up into bins and the probability of each bin containing a galaxy was calculated. The t bins were 0.5 Gyr in width and ranged from 0 and 15 Gyr, the [Z/H] bins were 0.125 dex in width and ranged from -2.2875 to 0.7125 dex, and the [α/Fe] bins were 0.025 dex in width and ranged from 0.0 to 0.5 dex. Monte Carlo simulations were used to estimate the probability of obtaining a likelihood statistic larger than the observed one by chance.

These simulations provide us with a great deal of information regarding the intrinsic scatter in each of the stellar population parameters. Firstly, it is difficult to constrain the e-folding of the exponential t distribution. This is due to small uncertainties in index strengths translating to large changes in the t estimate combined with relatively large observational errors in the Hβ index. Although models with small values of e-folding are not ruled out, we find that our data are most consistent with the model having $\tau = 900$ Myr. Secondly, small scatters in [Z/H] ($\sigma_{[Z/H]} < 0.1$ dex) are strongly ruled out, as are large scatters ($\sigma_{[Z/H]} > 2.0$ dex). We find the model with $\sigma_{[Z/H]} = 0.3$ dex is the most consistent with our data. Finally, large scatters in [α/Fe] ($\sigma_{[\alpha/Fe]} > 0.3$ dex) are also strongly ruled out. Models with low scatters in the [α/Fe] are not ruled out, but we find that the most consistent model is that with $\sigma_{[\alpha/Fe]} = 0.07$ dex.

As we showed above, both [Z/H] and [α/Fe] are significantly correlated with σ. Since the galaxies in the low-z cluster sample have a range of σ, it is understandable that we should detect an intrinsic scatter in [Z/H] and [α/Fe]. The degree to which the intrinsic scatter in these parameter distributions is attributable to trends with σ can be estimated by comparing the scatter expected given the σ distribution in the cluster and the intrinsic scatter about the parameter–σ relations. Since the correlation between t and σ is only marginal we will concentrate on the scatter in [Z/H] and [α/Fe].

We use the parameter–σ relations to convert the galaxies' σ into parameter values. The scatter in these values is then calculated; the intrinsic scatter about the relation is then added in quadrature to the scatter due to the relation with σ, and the result is compared to that obtained from the simulations described above.

The scatter expected in [Z/H] on this basis is 0.21 dex, which is comparable to the intrinsic scatter of $\sigma_{[Z/H]} \sim 0.3$ dex estimated above. The expected scatter in [α/Fe] is 0.06 dex, which is very close to the estimated intrinsic scatter $\sigma_{[\alpha/Fe]} \sim 0.07$ dex. It appears then that the intrinsic scatter in both the [Z/H] and [α/Fe] distributions can be accounted for by the parameter–σ relation and the intrinsic scatter about it.

Which of these two sources contributes most to the scatters is an interesting question. For the [Z/H] distribution, we find that the major contribution comes from the correlation with σ. For the [α/Fe] distribution, the major contribution comes from the intrinsic scatter about the [α/Fe]–σ relation. In analysing the stellar population parameter distributions in each of the four clusters, we found evidence of differences in the [α/Fe] distributions. Thus, we attribute the intrinsic scatter about the [α/Fe]–σ relation to differences in cluster properties (e.g. its dynamical state). No variation in the [Z/H] distributions between clusters was found, and so we attribute the scatter in the [Z/H]–σ relation to differences in galaxy properties.

1.6 Star formation inside out

One would expect that the projected radial distance of a galaxy from the cluster centre (R_{proj}) would be correlated with the local density: the cluster core being more

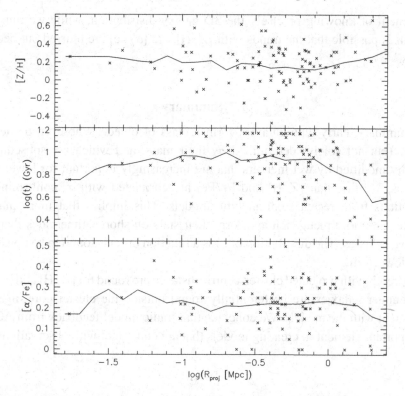

Figure 1.3 The trend of mean $[Z/H]$ (*top*), mean age (*middle*), and mean $[\alpha/Fe]$ (*bottom*) with projected cluster-centric radius. The mean is calculated in 0.4 dex wide bins at steps of 0.1 dex.

dense than the cluster outskirts, which in turn are more dense than the surrounding structures. Examining the stellar population parameters as functions of R_{proj} should therefore indicate variations with local density. Figure 1.3 shows the variations of each stellar population parameter with R_{proj}. The lines in these plots show the mean values in logarithmic distance bins 0.4 dex in width at 0.1 dex steps. Only galaxies inside the Abell radius (R_{Abell}) are shown, since outside this radius the galaxies are too sparsely sampled. From this figure we see that both $[Z/H]$ (top) and $[\alpha/Fe]$ (bottom) are essentially constant within R_{Abell}. On the other hand, t begins to decrease outside a radius of ~ 0.8 Mpc. Inside this radius $\langle t \rangle \approx 8$ Gyr, but this decreases to $\langle t \rangle \approx 4$ Gyr at R_{Abell}. This decrease is significant at the $> 4\sigma$ level. This result shows that galaxies outside the cores of clusters are more commonly younger or have had a more recent bout of star formation. These could possibly represent galaxies that have had a burst of star formation triggered by their movement into the cluster environment. These results confirm the De Lucia *et al.* (2006) prediction of t gradients within clusters. We cannot, however, confirm the predicted $[Z/H]$ and $[\alpha/Fe]$ radial gradients. It must be noted that the semi-analytical models have the benefit of knowing precisely the 3D galaxy distribution, while we must use R_{proj}. It is possible that the trends with $[Z/H]$ and $[\alpha/Fe]$ are blurred out because of this.

1.7 Summary

In summmary, early-type galaxies form a class of objects whose properties are largely (but not solely) determined by their mass, and which are subsequently partially modified by wet mergers that are increasingly important for lower mass galaxies. We find that $[Z/H]$ and $[\alpha/Fe]$ are correlated with σ, confirming the predictions from recent semi-analytic models. This implies that more massive galaxies are more metal-rich and form their stars on shorter timescales than less massive galaxies. The distribution of t (as a function of σ) is found to be consistent with down-sizing.

No trends with projected cluster-centric distance are found for $[Z/H]$ and $[\alpha/Fe]$, but younger galaxies are preferentially found outside the cluster core, again in agreement with the models. We conclude that 'radio mode' feedback from AGN is an important element in creating models that are consistent with observations.

References

Burstein, D., *et al.* (1984). *Astrophys. J.*, **287**, 586
Collobert, M., *et al.* (2006). *Mon. Not. R. Astron. Soc.*, **370**, 1213
Cowie, L. L., *et al.* (1996). *Astrophys. J.*, **112**, 839
Croton, D. J., *et al.* (2006). *Mon. Not. R. Astron. Soc.*, **365**, 11
De Lucia, G., *et al.* (2006). *Mon. Not. R. Astron. Soc.*, **366**, 499
Kauffmann, G. (1996). *Mon. Not. R. Astron. Soc.*, **281**, 487
Proctor, R. N., Forbes, D. A., & Beasley, M. A. (2004). *Mon. Not. R. Astron. Soc.*, **355**, 1327
Thomas, D., Maraston, C., & Bender, R. (2003). *Mon. Not. R. Astron. Soc.*, **339**, 897
Thomas, D., *et al.* (2005). *Astrophys. J.*, **621**, 673
Trager, S. C., *et al.* (1998). *Astrophys. J. Suppl.*, **116**, 1
Trager, S. C., *et al.* (2000). *Astrophys. J.*, **120**, 165

2

Suppressing cluster cooling flows by multiple AGN activity

Adi Nusser

2.1 Introduction

Models invoking only the central AGN to resolve the cooling flow conundrum in galaxy clusters require fine-tuning of highly uncertain microscopic transport properties to distribute the thermal energy over the entire cluster cooling core. A model in which the ICM is heated instead by multiple, spatially distributed AGNs bypasses most of these difficulties (Nusser *et al.* 2006). The central regions of galaxy clusters are rich in spheroidal systems, all of which are thought to host black holes and could participate in the heating of the ICM via AGN activity of varying strengths. And they do. There is mounting observational evidence for multiple AGNs in cluster environments. Active AGNs drive bubbles into the ICM. We identify three distinct interactions between the bubble and the ICM: (1) Upon injection, the bubbles expand rapidly *in situ* to reach pressure equilibrium with their surroundings, generating shocks and waves whose dissipation is the principal source of ICM heating. (2) Once inflated, the bubbles rise buoyantly at a rate determined by balance with the viscous drag force, which itself results in some additional heating. (3) Rising bubbles expand and compress their surroundings. This process is adiabatic and does not contribute to any additional heating; rather, the increased ICM density due to compression enhances cooling. Our model sidesteps the "transport" issue by relying on the spatially distributed galaxies to heat the cluster core. We include self-regulation in our model by linking AGN activity in a galaxy to cooling characteristics of the surrounding ICM. We use a spherically symmetric one-dimensional hydrodynamical code to carry out a preliminary study illustrating the efficacy of the model. Our self-regulating scenario predicts that there should be enhanced AGN activity of galaxies inside the cooling regions compared to galaxies

AGN Feedback in Galaxy Formation, eds. V. Antonuccio-Delogu and J. Silk. Published by Cambridge University Press. © Cambridge University Press 2011.

in the outer parts of the cluster. This prediction is sustained by recent observational results.

2.2 Outline of the model

Typically, studies investigating the impact of AGNs have tended to focus on AGN activity in only the central galaxy, and invoke relatively high values of thermal conduction to redistribute the energy across the cooling core (see, for example, Roychowdhury *et al.* 2005). Here, we propose instead that the cluster ICM is heated by the AGN activity in more than just the one central galaxy. The central region of the clusters is rich in giant elliptical galaxies, most of which are presumed to harbor supermassive black holes that are thought to inject radio bubbles into the ICM with a duty cycle of $\sim 10^8$ yr per Hubble time.

Upon injection, the overpressurized bubbles rapidly expand (virtually *in situ*) to reach pressure equilibrium with their surroundings, generating waves and shocks, the dissipation of which is the primary mechanism by which the ICM is heated. Once in pressure equilibrium, the bubbles will rise buoyantly. The dissipation of the work done by viscous drag acting to retard the bubbles' motion will further heat the ICM, though to a much lesser extent. In response to the decreasing pressure, the rising bubbles will expand. This expansion causes a corresponding compression of the ICM. However, while the initial inflation of the cavity may trigger internal gravity waves, compression waves and shocks that heat the ICM, this subsequent expansion–compression is adiabatic in character and no heating of the ICM ensues, even though the ICM's internal energy will increase due to compression. In fact, the compression actually leads to the erosion of the thermal content of the ICM because the resulting increase in density results in enhanced cooling. This can be seen as follows. The compression of the ICM can be approximated as an adiabatic process so that the "entropy" $S = T/n^{2/3}$ is conserved. The bremsstrahlung cooling time is $t_c \propto T/(nT^{0.6}) \sim S^{0.4} n^{-1.06}$. Therefore, cooling becomes more efficient as n is increased during adiabatic compression.

In our model, the heating occurs near where the bubbles are produced but since the galaxies that produce these bubbles are distributed throughout the cluster core region, so too is the heating. Modest heat conduction should suffice to distribute the energy further over the ICM between the galaxies. Since the energy deposition is a local phenonemon, it is relatively straightforward to construct a self-regulating model.

To model the interaction of the bubbles with the ICM (which we treat as all material not belonging to the hot and dilute bubbles), we use a spherical one-dimensional (1D) hydrodynamical code. The code treats the bubbles and the ICM as a two-fluid system in a fixed dark matter halo. The code treats the interaction of the bubbles and ICM in a consistent semi-analytical way and assumes that the distribution of bubbles is spherically symmetric. The code integrates the hydro-dynamical equations of an *ambient medium* with thermodynamic properties that depend on the local filling factor of bubbles.

2.2.1 A detailed description

A 1D hydrodynamical code cannot tackle the intrinsically three-dimensional (3D) problem of the evolution of individual bubbles in the ICM, in particular if bubbles are injected from sources away from the center as in our proposed scenario. Therefore, a formalism for following the evolution of bubbles statistically must be developed. We represent bubbles as a fluid specified by the following physical quantities: the radius of bubbles, their number density, and velocity, all as a function of distance from the center. The bubbles and the ICM are a two-fluid system that is best described as a single fluid, which we term the *ambient medium*. The representation in terms of a single medium is applicable only if local pressure equilibrium between the bubbles and the ICM is established. Hereafter, quantities related to the ambient medium, the bubbles, and the ICM (gas outside bubbles) are denoted by the subscripts, a, b, and I, respectively.

In our model, cluster galaxies affect the ICM by the production of over-pressurized bubbles made of hot relativistic plasma with an adiabatic index $\gamma_b = 4/3$. Bubbles are produced in response to local cooling in the vicinity of galaxies according to the self-regulating mechanism described below.

Let $n_b(r)$, $v_b(r)$, and $R_b(r)$, be, respectively, the number density of bubbles, the velocity of bubbles relative to the ambient medium, and the radius of bubbles, all at distance r from the center. The motion of the bubbles is modeled as that of a pressure-free (i.e. collisionless) fluid with a velocity that is determined by a balance between buoyancy and drag forces with the ICM. Bubbles are assumed to be injected with the same energy per bubble and the same initial pressure so that at distance r all bubbles have the same size (see Equation 2.1) and the same velocity. Therefore, a hydrodynamical description for the flow of bubbles is self-consistent. The pressure-free assumption is also self-consistent since $v_b(r)$ is a single-valued function of r so that no "shocks" can form.

The various phases of the evolution of the bubbles are as follows. *Phase I*: A bubble is injected at distance r from the center. This over-pressurized bubble undergoes a supersonic expansion in the ICM until its internal pressure drops to close to that of the ICM. During this phase of rapid expansion, the bubbles heat the ICM by the production of weak shocks. This expansion is assumed to be quite rapid and such that the bubble does not move significantly away from the distance at which it has been injected. *Phase II*: The dilute hot bubble then

rises by buoyancy towards more distant regions of lower pressure. The moving
bubbles heat the ICM via viscous drag forces. As noted, the velocity of bubbles
relative to the ambient medium is determined by a balance between buoyancy and
drag forces with the ICM. Rising bubbles also expand adiabatically, compressing the surrounding ICM. This process does not heat the ICM. Increased ICM
density does, however, elevate the efficiency of cooling. *Phase III*: Finally, bubbles are destroyed by hydrodynamical (Rayleigh–Taylor and Kelvin–Helmholtz)
instabilities.

Let a bubble be injected with an initial energy E_{bi} and an initial internal pressure
p_{bi}. Since $E_{bi} = u_{bi}\rho_{bi}4\pi R_{bi}{}^3/3$, where ρ_{bi} and u_{bi} are the initial mass density and
energy per unit mass in the bubble, the equation of state $p_b = (\gamma_b - 1)\rho_b u_b$ yields,
$R_{bi} = \left[\frac{3}{4\pi}(\gamma_b - 1)\frac{E_{bi}}{p_{bi}}\right]^{1/3}$. The equation of state of the material inside the bubble
gives $E_b = (4\pi/3)R_b{}^3 p_b/(\gamma_b - 1)$ for the thermal energy of a bubble of radius R_b
with internal pressure p_b. At the end of the rapid expansion in *Phase I* we assume
that $p_b \approx p(r)$ where $p(r)$ is the ICM pressure at the distance at which the bubble
has been injected. We find heat transferred to the ICM by the generation of shocks
is

$$\Delta E_1 = C_{wsh}(E_{bi} - E_{b0}) = C_{wsh}E_{bi}\left[1 - \left(\frac{p}{p_{bi}}\right)^{1-\frac{1}{\gamma_b}}\right], \qquad (2.1)$$

where $0 \le C_{wsh} \le 1$ is an efficiency parameter. In the present treatment, we assume
that the energy transferred to the ICM is deposited locally.

The dynamical evolution of the system comprising an expanding, rising bubble
in a bubbly medium subject to radiative cooling is, in general, a highly non-trivial
problem, involving a host of complex phenomena including the compressible nature
of the bubbly medium, the evolution and dissipation of shocks, wakes, and waves
in this medium, bubble–bubble interactions mediated by bubble wakes, bubble
coalescence, changing bubble shapes and orientations, etc. We write the equations
of motion of a bubble in *Phase II*. We restrict ourselves to a simple situation of a
system involving a single spherical bubble rising along the radial direction in the
ICM. A bubble of radius R_b at distance r experiences a buoyant force of $F_{bu} = \frac{4\pi}{3}R_b^3\rho_a g_{eff}$, where we have assumed that $\rho_b \ll \rho_a$. Here, g_{eff} is the gravitational force
field (per unit mass) as measured in a frame of reference attached to the ambient
medium at distance r. The actual gravitational field, g, differs from g_{eff} only when
the system deviates from hydrostatic equilibrium, such as in a cooling runaway
where $g_{eff} \approx 0$. We write the drag term on the bubble as it moves through the ICM
with velocity v_b as $F_{drag} = \frac{\pi}{2}C_d R_b^2\rho_a v_b^2$, where C_d is the drag coefficient. The rise
velocity of the bubbles can be estimated by equating the drag and buoyancy forces
yielding $v_b^2 = 8g_{eff}R_b/3C_d$. As a bubble rises, the gravitational potential energy of

the system (the bubble and the surrounding medium) decreases. A rising bubble also expands and its internal energy decreases. The total energy available as the bubble rises from position 1 (point of injection) to position 2 is $\int_1^2 \frac{4\pi}{3} R_b^3 \rho_a \frac{\partial \phi}{\partial r} dr = \gamma_b(E_{b1} - E_{b2})$, where ϕ is the gravitational potential. Consequently, $\Delta E_2 = (1 + \gamma_b)(E_{b1} - E_{b2})$. This is the energy available to lift the bubble against the drag force, overcome the inertia of displacing the surrounding fluid, as well as compress the ambient medium. In general, it is not possible to determine how much of this energy will go towards compressing the ambient medium and how much will go towards heating the ICM, be it due to drag or through the eventual dissipation of the wakes and motions associated with the displacement of the ICM, unless one knows the details of the rise–expansion process. In the limit that the bubble rises slowly, maintaining pressure equilibrium with its surroundings at all times, the bubble expansion/ICM compression is adiabatic and the change in the internal energy of the bubble goes entirely towards raising the internal energy of the ambient medium. In this case, the total energy available for heating the ICM is simply that due to the change in the potential energy and the corresponding heating rate (in units of energy per unit time) is $v_b F_{bu} = (4\pi/3)R_b^3 \rho_a g_{eff} v_b$. The heating rate is $\mathcal{H}_{heat} = \frac{4\pi}{3} R_b^3 n_b g_{eff} v_b$. It is instructive to relate this heating term to the physical parameters of the bubbles and their injection rate. Assume that bubbles are injected within a sphere of radius R_{inj} around the center, at a rate $\dot{N}_{inj}$ per unit volume. Neglecting bubble destruction, the total number of bubbles within a distance $r \gg R_{inj}$ is $\dot{N}_{inj}(4\pi/3)R_{inj}^3 r/v_b$, so that the number density within r is $\bar{n}_b(< r) \approx \dot{N}_{inj} R_{inj}^3/(v_b r^2)$. Approximating $n_b(r) \sim \bar{n}_b(r)$ in (2.1) gives

$$\mathcal{H}_{heat} \sim \frac{4\pi}{3} \dot{N}_{inj} R_{inj}^3 g_{eff} \frac{R_b^3}{r^2} . \tag{2.2}$$

It is interesting that this does not depend on the drag coefficient, C_d. We can further simplify this expression by noting that the bubble energy at r is $E_b \sim 4\pi R_b^3 p/3(\gamma_b - 1)$. The final result is

$$\mathcal{H}_{heat} \sim (\gamma_b - 1)\dot{N}_{inj} g_{eff} \frac{R_{inj}^3}{r^2} \frac{E_b}{p} . \tag{2.3}$$

For $r < R_{inj}$ the heating rate is obtained by replacing R_{inj} with r in the last expression.

We write the energy equation in terms of the ICM entropy $S_I = u_I/\rho_I^{\gamma_1-1}$ as

$$\rho_I^{\gamma_1-1}\dot{S}_I = \mathcal{H} - \frac{u}{t_{cool}} , \tag{2.4}$$

where t_{cool} is the cooling time. Two sources contribute to $\mathcal{H}$: (1) heating by weak shocks generated during the initial rapid expansion of the over-pressurized bubbles, and (2) heating by drag with the ICM as bubbles rise up by buoyancy. Adiabatic work done by the bubbles as they move upward does not change S_I and therefore does not contribute to $\mathcal{H}$. We introduce R_{inj}, the distance from the center within which bubbles are injected. At $r > R_{inj}$, bubbles can be present only as a result of flow of bubbles from regions with $r < R_{inj}$. Given the injection rate (number per unit volume per unit time), $\dot{N}_{inj}(t, r)$, of bubbles, we write the local heating rate (energy per unit mass of the ICM) via weak shocks as

$$\mathcal{H}_{wsh} = \dot{N}_{inj} \frac{\Delta E_1}{\rho_a} , \qquad (2.5)$$

where ΔE_1 is given by (2.1). As a recipe for self-regulated feedback, we assume that bubbles are generated only if $\dot{S}_I$ is negative. We write the flux of injected bubbles at a point $r \le R_{inj}$ as

$$\dot{N}_{inj} = -\frac{\eta}{E_{bi}} \frac{\dot{S}_I}{S_I} \rho_I c^2 \quad \text{for} \quad \dot{S}_I < 0 , \qquad (2.6)$$

and zero, otherwise. The free parameter η represents the product of the mass fraction of cold gas that accretes onto the AGNs and the efficiency of AGNs at transforming the accreted mass into bubbles. The cold gas is assumed to be generated at a rate (in units of mass per unit volume per unit time) of $\rho_I \dot{S}_I / S_I$. This cold gas is assumed to form by condensation via the development of thermal instabilities (e.g. Field 1965). Finally, c is the speed of light and, as before, E_{bi} is the energy with which a bubble is injected.

We can use the above formalism to investigate the role of heating by weak shocks, the dominant source of heating in our model. Substituting ΔE_1 and $\dot{N}_{inj}$ from (2.1) and (2.6), respectively, into (2.5) yields

$$\mathcal{H} = -u_A \frac{\dot{S}_I}{S_I} , \quad \text{where} \quad u_A = C_{wsh} \eta c^2 \left[1 - \left(\frac{p}{p_b} \right)^{1 - \frac{1}{\gamma_b}} \right] , \qquad (2.7)$$

where $p = p(r)$ is the ambient pressure at r and p_b is the internal bubble pressure with which the bubbles are injected. This expression holds for $\dot{S}_I < 0$. For $\dot{S}_I > 0$, we have $\mathcal{H} = 0$. Substituting (2.7) in (2.4) and solving for $\dot{S}_I$ we get,

$$\dot{S}_I = -\frac{S_I}{\tau_{cool}} , \quad \text{where} \quad \tau_{cool} = \left(1 + \frac{u_A}{u} \right) t_{cool} . \qquad (2.8)$$

This equation implies that cooling is substantially suppressed at low temperatures, $u \ll u_A$, while at higher temperatures, shocks are unable to suppress cooling. For an ICM at a temperature corresponding to velocity V, an efficient suppression of cooling at distance r occurs if $C_{wsh}\eta \sim \left(\frac{V}{c}\right)^2 \approx 10^{-5}\left(V/10^3 \text{ km s}^{-1}\right)^2$, where we have assumed $p_b \gg p$. We develop a numerical hydrodynamical model to describe the evolution of the ICM and the bubbles in a fixed dark matter halo. The density profile in the halo is approximated by the NFW profile of the form $\rho(s,t) \propto 1/s(1+cs)^2$, where $s = r/R_v$ and R_v is the virial radius of the halo. We adopt the value $c = 5$ for the concentration parameter. The ambient medium is represented by lagrangian shells of fixed mass. In each time-step, the equations of motion of the bubble and the ambient fluids are solved in the following order.

2.3 Results

The initial temperature profile is assumed to be isothermal with a temperature, T (in keV), given by $1.16 \times 10^7 k_B T/\mu m_p = (2/3)GM_v/R_v$ where k_B is the Boltzmann constant, μ is the mean molecular weight and m_p is the mass of the proton. The equation for hydrostatic equilibrium implies a slope of $-9/4$ for the gas density at $r = R_v$. The gas density at any r $(r < R_v)$ is solved numerically using the equation of hydrostatic equilibrium including gas gravity. In all runs, the code evolves the shells from redshift $z = 2$ until $z = 0$. In some runs with insufficient feedback to counter cooling, the hydrodynamic time-step becomes exceedingly small. Whenever this happens, the shells colder than 1 keV and denser than 1.5 cm^{-3} are held fixed and are not moved by the code at any later time. Results from simulations for clusters with a virial radius of $R_v = 4$ Mpc are shown in Figure 2.1. In the column to the left we show results that include radiative cooling and no heating. Effects of cooling become significant by $z = 1$. At $z < 0.5$ a cooling runaway develops in the inner regions: cooling becomes increasingly more rapid due to the enhanced density and lower temperature.

The effect of heating by bubbles, according to our feedback recipe described in Section 2.2.1, is explored in the middle and right panels for two values of the efficiency parameter η. To model the effect of multiple AGN activity, bubbles are allowed to be injected anywhere in the ICM (i.e. $R_{inj} = R_v$). As noted before, the heating rate of the ICM does not depend on the initial bubble energy, E_{bi}. All results in these two columns are for $C_d = 1$ and $C_{wsh} = 1$, and an initial bubble pressure of 10^3 times the central ICM pressure. Cooling flows for the two different clusters are completely suppressed for η as low as 0.01, while a value $\eta = 10^{-5}$ gives moderate suppression of cooling as is required by the observations.

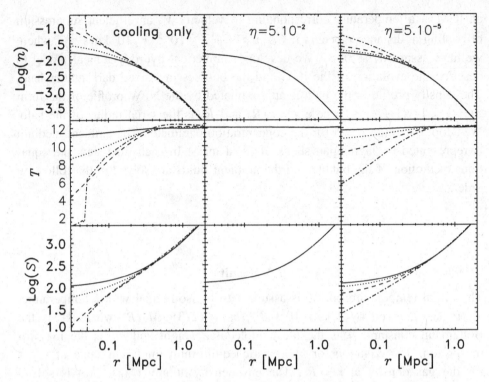

Figure 2.1 The ICM electron number density (top), temperature (middle), and entropy (bottom) profiles versus distance from the cluster center (in Mpc). The left column shows profiles obtained with radiative cooling and no energetic feedback. In the middle and right panels feedback is done via bubbles generated in the entire cluster with efficiency parameters $\eta = 10^{-2}$ and $\eta = 10^{-5}$, respectively. The solid, dotted, dashed, and dash-dotted curves correspond to redshifts $z = 1.6$, 1, 0.5, and 0, respectively. The results correspond to an NFW dark halo profile with a virial radius of $R_v = 4$ Mpc and a concentration parameter of $c = 5$. The number density, temperature, and entropy are given in cm^{-3}, keV, and keV cm^2, respectively.

2.4 Summary and discussion

We propose that modest radio outbursts from many cluster galaxies may help to quench cluster cooling flows. We appeal to typical massive elliptical galaxies that are thought to have radio outbursts with a duty cycle of $\sim 10^8$ yr per Hubble time and are concentrated in the cluster core (within the central 200 kpc or of order twice the extent of the cooling region). Weak shocks generated during the initial bubble expansion and viscous/turbulent drag on buoyantly rising bubbles are the principal dissipative heating processes. Indirect evidence for multiple heating sources in the form of AGNs comes from the frequent presence of X-ray "ghost" cavities that survive long after any associated radio lobes have decayed, and from amorphous radio halos that further suggest an ICM percolated with a distribution of bubbles whose contrast in the X-ray has been diminished due to adiabatic expansion (cf. Heinz & Churazov 2005). Our scenario is in agreement with recent observational evidence for the existence of AGN activity in about 5% of cluster galaxies at low redshift (Martini *et al.* 2006). Cosmological evolution of the AGN inevitably increases the feedback at earlier epochs ($z \lesssim 2$). Our scenario alleviates some of the problems that may arise if energy feedback is generated by a single central source (AGN) harbored in the central cD galaxy. The energy transport needed over the entire cooling flow region is not easy to provide if the energy source is associated only with the central AGN (e.g. Mathews *et al.* 2005). One approach is to ensure that mechanical disturbances are able to propagate large distances before they are damped by viscosity (e.g. Ruszkowski *et al.* 2004; Reynolds *et al.* 2005). This approach has two potential shortcomings. Firstly, the heating rate per baryon of dissipating waves should decay with distance from the central source as r^{-3}. This is too steep a gradient for the heating to affect the ICM over the entire region that is susceptible to cooling. In fact, Roychowdhury *et al.* (2004) find that the heating leads to a convectively unstable core profile that would potentially destroy any pre-existing metallicity gradient, in conflict with the X-ray observations. Secondly, the viscosity has to be finely tuned: too high a value results in dissipation much too closely concentrated near the central source, and too low a value means the heating is inefficient. Since the relevant viscosity coefficient scales as $T^{5/2}$, it is difficult to envision that a viscous solution that relies on fine-tuning prevails in all clusters. Moreover, the ICM is weakly magnetized, and the nature of transport processes, such as thermal conduction and viscosity, in such media is not well understood.

The model presented here heats the ICM through shocks generated during the initial inflation of the bubbles, and drag forces acting on the rising bubbles. However, the latter rely on some viscosity to dissipate heat but this is not an important source of heating and, with respect to the former, we do not need to fine-tune the viscosity since the energy distribution is affected by multiple AGNs hosted in a spatially extended distribution of galaxies. Moreover, it may be possible to avoid the issue altogether. Heinz and Churazov (2005) have recently argued that in a bubble-filled medium of the kind proposed here, the bubbles themselves act as catalysts to convert shock waves into heat. Additionally, the interaction between the bubbles and the strong shocks will both broaden and distort the shock fronts, causing the features to appear weaker in X-ray observations and resulting in mistaken assumptions that cavity inflation is a gentle process and that the ICM is not subject to strong shocks. Both of these features further contribute to making our model highly viable.

An additional principal feature of our model is the incorporation of self-regulation between heating and cooling. This should inevitably lead to a correlation between the mechanical heating luminosity and the cooling rate. The former is measured by the properties of X-ray cavities and the latter by the cluster X-ray luminosity. Evidence for such a correlation is given by Bîrzan *et al.* (2004).

In fact, the estimated ratio of mechanical to X-ray luminosity in this latter paper falls short of that needed to quench cooling flows on group and cluster scales by a factor of 3–10. However, the inferred mechanical luminosities are based on cavity pV estimates. In our model, a primary source of energy transfer to the ICM is via weak shocks generated by the expansion of the initially over-pressurized bubble. The initial bubble energy is related to the cavity pV by Equation 2.3, and the resulting injected mechanical energy may exceed pV by a factor of up to 10. Note that this equation neglects the work done by the AGN jet to place the initial seed bubble in the ICM, which can be dissipative and contribute to heating the ICM if the seed bubble is ejected with supersonic speeds. Furthermore, these estimates neglect the energy content in waves detected in a few clusters. Our model consequently helps in resolving the apparent discrepancy in the estimated feedback required to halt cooling flows even in massive clusters.

References

Bîrzan L., Rafferty D. A., McNamara B. R., Wise M.W., Nulsen P. E. J., 2004, *ApJ*, **607**, 800
Field G. B., 1965, *ApJ*, **142**, 531
Heinz S., Churazov E., 2005, astro-ph/0507038
Martini P., Kelson D. D., Kim E., Mulchaey J. S., Athey A. A., 2006, astro-ph/0602496
Mathews W. G., Faltenbacher A., Brighenti F. 2005, astro-ph/0511151
Reynolds, C. S., McKernan B., Fabian A. C., Stone J. M., Vernaleo J. C., 2005, *MNRAS*, **357**, 242
Roychowdhury S., Ruszkowski M., Nath B. B., Begelman, M. C., 2004, *ApJ*, **615**, 681
Roychowdhury S., Ruszkowski M., Nath B. B., 2005, *ApJ*, **634**, 90
Ruszkowski M., Brüggen M., Begelman M.C., 2004, *ApJ*, **611**, 158

3

Starburst and AGN activity in Spitzer-selected sources at high-z

M. Polletta, A. Omont, C. Lonsdale & D. Shupe

3.1 Introduction

Some galaxy evolutionary models postulate that powerful starburst galaxies at high-z yield local massive galaxies following the effects induced by an accreting supermassive black hole (SMBH) at their centre (e.g. Di Matteo *et al.* 2005). However, it is not clear on which spatial and temporal scales and through which physical processes this transition takes place (see Coppin *et al.* 2008). Here, we investigate this evolutionary scenario by comparing star formation rates (SFRs), AGN activity and stellar masses in high-z ($z \sim 2$) active systems.

3.2 Spitzer selection of high-z luminous infrared galaxies

For this work, we selected a sample of IR luminous source candidates in a $\sim 20 \deg^2$ area obtained by combining the Lockman Hole field (LH, $\sim 11 \deg^2$, $\alpha = 10^h 45^m$, $\delta = +58°$), and the XMM-LSS field (XMM, $\sim 9 \deg^2$, $\alpha = 02^h 21^m$, $\delta = -04° 30'$) of the Spitzer Wide Area Infrared Extragalactic Survey (SWIRE[1]; Lonsdale *et al.* 2003). Both fields benefit from multi-band ground-based optical (Ugriz) and Spitzer IR bands (seven bands from 3.6 to 160 μm). IR luminous sources, powered by star formation or AGN activity, are expected to be bright mid-infrared (MIR) sources. Powerful starburst galaxies are characterised by spectral energy distributions (SEDs) that are bright throughout the MIR to millimetre range. Luminous AGNs are bright MIR sources because their emission from AGN-heated dust peaks in the MIR. We thus selected all sources with a 24 μm flux > 400 μJy (corresponding to $\gtrsim 5\sigma$). We then divided the sample into three classes: starbursts, AGNs and

[1] http://swire.ipac.caltech.edu/swire/swire.html

AGN Feedback in Galaxy Formation, eds. V. Antonuccio-Delogu and J. Silk. Published by Cambridge University Press. © Cambridge University Press 2011.

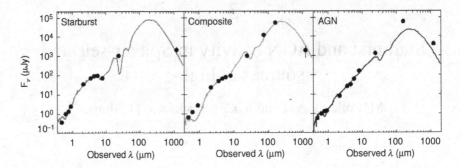

Figure 3.1 Examples of spectral energy distributions (full circles) and best-fit templates (solid line) of a starburst (*left panel*), a composite system (*middle panel*), and an AGN (*right panel*) from the selected sample.

composite systems, based on their SEDs. The starburst selection requires an optical r'-band magnitude $r' > 23$ to favour high-z systems, and a maximum at 5.8 μm out of all four IRAC fluxes (i.e. $f_{3.6} < f_{4.5} < f_{5.8} > f_{8.0}$). This maximum is interpreted as being the peak of stellar light ($\lambda \sim 1.6$ μm rest-frame) when redshifted to $z \sim 2.2$–3.2. The selection of luminous obscured AGNs and composite objects (AGN + starburst) at high-z requires red optical-IR colours ($r' - [3.6] > 6.5$) to favour high-z sources, and MIR SEDs indicative of AGN-heated hot dust emission. In particular, the AGNs are those that exhibit red ($F_\nu \propto \nu^{-2}$) and smooth MIR SEDs, while composite (AGN + starburst) objects show a red MIR SED on top of starlight emission. Examples of optical-MIR SEDs of each class are shown in Figure 3.1, and their MIR colours are illustrated in Figure 3.2. AGNs show the reddest colours typical of hot dust emission (see QSO2 template) and the starbursts cluster near the high-z end of the starburst and ULIRG templates (Figure 3.2). Composite objects show intermediate colours, but they tend to overlap more with the starbursts as their AGN component starts dominating at longer wavelengths, i.e. at $\lambda \geq 8$ μm. Each source SED (from U to 24 μm) is fitted using the Hyper-z code (Bolzonella *et al.* 2000) and a library of galaxy and AGN templates in order to estimate its photometric redshift and expected far-infrared (FIR) flux following the same procedure described in Lonsdale *et al.* (2009). Typical uncertainties on the photometric redshift are ±0.5. The sources with the brightest predicted FIR fluxes and $z \gtrsim 2$ made our final sample. Long wavelength data (far-IR and mm/sub-mm) are necessary to confirm the large IR luminosities. We have thus undertaken a systematic observing programme at the IRAM 30 m telescope using the MAMBO array operating at 1.2 mm to probe the FIR luminosities of our IR luminous candidates. Of our best candidates, we were able to observe with MAMBO 65 sources: 33 starbursts, 22 obscured AGNs, and 10 composite systems. The results on the starburst galaxies from this programme have been presented in Lonsdale *et al.* (2009).

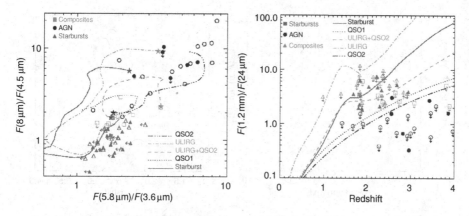

Figure 3.2 *Left panel:* $F(5.8\,\mu m)/F(3.6\,\mu m)$ as a function of $F(8.0\,\mu m)/$ $F(4.5\,\mu m)$ for the selected targets: AGNs (circles), starbursts (triangles), and composites (squares). Filled symbols represent sources with $F(1.2\,mm) \geq 2\sigma$, and open symbols those with $F(1.2\,mm) < 2\sigma$. The curves represent the expected shades as a function of redshift ($0 < z < 2$, with $z = 0$ marked by a star symbol) for various templates from Polletta *et al.* (2007), QSO2: SWIRE J104409.95+585224.8; ULIRG: IRAS 22491−1808; ULIRG+QSO2: IRAS 19254−7245 S; QSO1: TQSO1; Starburst: M 82. *Right panel:* $F(1.2\,mm)/$ $F(24\,\mu m)$ as a function of redshift. Same symbols as in the left panel.

Here, we present the entire sample and a comparison between the starbursts and the AGNs.

3.3 MAMBO observations and results

MAMBO observations were carried out in 2005–6 in standard on–off photometry observing mode and with ∼1 hr exposure per source (for more details see Lonsdale *et al.* (2009)). The 24 μm and 1.2 mm fluxes and the redshifts of all selected sources are shown in Figure 3.3. The percentage of sources with a 1.2 mm flux >2σ is 27% for the AGN sample, 39% for the starburst sample, and 10% for the composite systems. The small detection fraction in the composite sample is attributed to the shorter exposure times of the MAMBO observations. The ranges of 1.2 mm fluxes measured for the three sub-samples are very similar, from 1.5 to 6 mJy, and the detection rate is independent of the 24 μm flux, and of the redshift. The AGNs are systematically brighter (>1 mJy) at 24 μm.

The 1.2 mm/24 μm flux ratios of the selected sources are shown in Figure 3.2. The sources in the diagram segregate according to their classification, with decreasing ratios going from starbursts to composites, and to AGNs. The AGNs show the lowest 1.2 mm/24 μm flux ratios in the whole sample. Such small ratios are attributed to

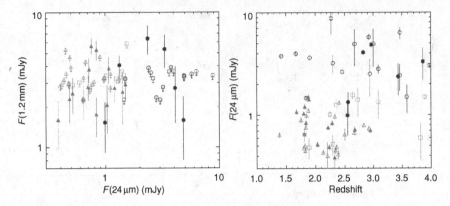

Figure 3.3 *Left panel:* 1.2 mm fluxes as a function of 24 μm fluxes. *Right panel:* 24 μm fluxes as a function of redshifts. Symbols as in Figure 3.2.

the contribution from AGN-heated hot dust emission to the 24 μm flux, rather than to a deficiency of flux at 1.2 mm.

3.4 AGN and starburst MIR emission contributions

Based on the analysis reported in Lonsdale *et al.* (2009), at least 30% of the selected starbursts contain an AGN, and the AGN contribution to the MIR emission is, on average, 34%. Based on the fraction of AGNs with $>2\sigma$ 1.2 mm fluxes (27%), we can claim that at least 27% of obscured QSOs host starburst activity. It is interesting to note that a large fraction of AGNs is detected at 70 μm (9 out of 22 sources or 41%), and that this is much larger than for the starbursts, 1 out of 33 sources. This suggests that the 70 μm emission in the AGN sample is mainly due to the contribution of the AGN component, as at 24 μm. Thus, the AGN-heated dust also emits significantly at 70 μm (rest-frame wavelength of 23 μm at $z = 2$), consistent with the vast majority of models that predict AGN-heated circumnuclear dust spectra.

In conclusion, we find that ∼30% of starbursts host AGN activity, and 27% of obscured AGNs are experiencing starburst activity. In the starbursts the AGN contributes to 34% of the MIR emission and much less in the FIR and the optical. In AGNs, the starburst dominates in the mm, while its contribution is negligible in the MIR up to 70 μm.

3.4.1 *Relative AGN and starburst timescales*

To estimate the relative timescales of the AGN and starburst phases, we consider the surface density of the two types of objects. The surface density of starburst-dominated objects as selected here (see Section 3.2) is about 125 deg^{-2}, while for luminous obscured AGNs with $F(24\,\mu m) > 1\,mJy$ it is about $33\,deg^{-2}$. Since at least 27% ($\sim 9\,deg^{-2}$) of all luminous and obscured AGNs are hosted by a starburst galaxy, the luminous AGN phase lasts $\sim 7\%$ [9/(125 + 9)] of the duration of the starburst phase. Based on the presence of moderate AGN activity in 30% of the starbursts, we estimate that obscured AGN activity, both moderately and highly luminous, is present during 37% (30 + 7) of the starburst lifetime. In other words, the starburst lifetime is almost three times longer than the AGN lifetime. Note that the lifetime of a sub-mm galaxy is estimated to be five to six times longer than the lifetime of a sub-mm-detected unobscured QSO (see Coppin *et al.* 2008 and references therein). This comparison implies that the obscured AGN phase lasts longer than the unobscured luminous QSO phase. Note that these surface densities are based on the initial SWIRE-selection, before optimising it for MAMBO observations. However, the relative fractions of starburst and AGN contribution are based on the SWIRE/MAMBO sample and are assumed to be valid for the whole SWIRE sample. This assumption is supported by the similarity in the average FIR-mm properties of a complete starburst sample (Fiolet *et al.* 2009) and of the MAMBO-optimised starburst sample selected here (Lonsdale *et al.* 2009).

3.5 Host galaxies

In Figure 3.4, we show the rest-frame luminosities (νL_ν) at $1.6\,\mu m$, and at $6\,\mu m$ as a function of the $1\,\mu m$ luminosity of all selected sources. Assuming a fixed mass-to-light ratio, $L(1\,\mu m)$, and $L(1.6\,\mu m)$ can be considered a proxy of stellar mass in starburst galaxies and composite systems. Since the AGN is highly obscured, $L(1\,\mu m)$ suffers less AGN contamination and it is a better tracer of stellar mass than $L(1.6\,\mu m)$ in AGNs. The AGN contribution makes the AGN systematically more luminous at 1 and $1.6\,\mu m$ than the starbursts and the composite systems at the same redshift (see Figure 3.4). At $1\,\mu m$ the three sub-samples exhibit the same range of luminosities, from 10^{11} to $2 \times 10^{12}\,L_\odot$. These luminosities correspond to stellar masses of $(0.8–16)\times 10^{11}\,M_\odot$. However, any AGN contribution at $1\,\mu m$ in the AGN sample would result in overestimated stellar masses; thus, it is probable that AGN hosts are less massive than starburst galaxies. The similar distribution of $L(1\,\mu m)$ for sources with different $1.2\,mm$ signal-to-noise, or with different classification, indicates that the estimated stellar masses are not related to the mm brightness or to the source classification.

The $6\,\mu m$ luminosity is a good proxy of AGN power in AGNs and composite systems because it is dominated by AGN-heated hot dust emission. In the starbursts, the AGN contribution at $6\,\mu m$ is absent or significantly lower than the emission (continuum and PAHs) associated with star-forming regions. Figure 3.4 clearly

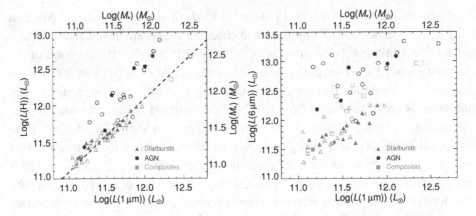

Figure 3.4 Monochromatic luminosity at rest-frame 1.6 μm (H-band) (*left panel*)
and at 6.0 μm (*right panel*) as a function of the monochromatic luminosity at 1 μm.
Symbols as in Figure 3.2. The dashed line represents the average ratio between
$L(1.6\,\mu m)$ and $L(1\,\mu m)$ for the starburst sample, or the typical ratio for the stellar
populations of this type of sources.

shows that the emission from hot dust results in a larger 6 μm luminosity in
composite systems and, even more, in AGNs. Starburst are systematically less
luminous at 6 μm, even if the sample includes some of the brightest mid-IR sources
in a 20 deg^2 area. Thus, the most powerful mid-IR sources are AGNs, and they are
about 10 times more luminous than the most luminous starbursts at the same
redshift at 6 μm. Similar results have been found at 8 μm from a large sample of
Spitzer-selected dusty galaxies (Weedman & Houck 2009).

In order to estimate the SFR of our sources we estimate the FIR luminosity
and apply the relationship found in local starburst galaxies (Kennicutt 1998). The
FIR luminosities are estimated by normalising a starburst template to the observed
1.2 mm flux or 3σ upper limit and integrating in the 8–1000 μm wavelength range.
In the case of upper limits, the luminosities and SFRs are also considered upper limits. The FIR luminosities and SFRs as a function of the 6 μm luminosities are shown
in Figure 3.5. The FIR luminosities range from $5 \times 10^{12}\,L_\odot$ to $3 \times 10^{13}\,L_\odot$, and the
corresponding SFRs from 800 to $6000\,M_\odot/\text{yr}$. Mm-detected AGNs and starbursts
show a similar range of SFRs, independently of their MIR (6 μm) luminosity.

3.6 Average SEDs

The average observed SEDs of the sources in the three groups are compared
in the left panel of Figure 3.5. The average SEDs are obtained by averaging
the observed fluxes in each band. The only exceptions are the fluxes at 70 and
160 μm that are instead measured from the stacked images following the same

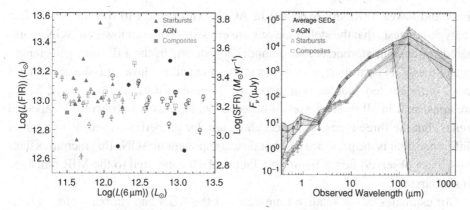

Figure 3.5 *Left panel:* FIR (8–1000 μm) luminosity as a function of the 6 μm luminosity. Symbols as in Figure 3.2. Downward arrows represent upper limits to the FIR luminosity. *Right panel:* Average observed SEDs of AGNs (filled circles), starbursts (triangles), and composite systems (squares). The average 70 and 160 μm fluxes have been measured in the stacked images. The average fluxes in the five optical bands and in the K-band are upper limits to the average flux because they are obtained using only the detected sources. The shaded area corresponds to the standard deviation of the mean. The mean values in the available bands have been connected with straight lines, but the underlying SED might be different, especially between the 8 and the 24 μm data points.

procedure as in Lonsdale *et al.* (2009). In the optical and in the K-band the average fluxes are obtained only from the detected sources, which are a minority in each group, therefore these average fluxes should be considered upper limits to the true averages. The comparison shows that the AGNs have a strong MIR component that can dominate the emission of a starburst like those selected here from 3.6 μm to 70 μm (1.2–23 μm in the rest-frame at $z = 2$). At 160 μm they are instead, on average, fainter than the starburst galaxies. At 1.2 mm, the average mm fluxes of all three classes are consistent. However, AGNs exhibit a wider range of mm fluxes than the other classes, as shown by the large uncertainties in Figure 3.5. This result emphasises the necessity of sampling the FIR emission of these objects at multiple wavelengths to estimate accurate FIR luminosities and dust temperatures.

3.7 Summary and conclusions

In this work, we investigated the postulated evolutionary link between high-z starburst galaxies, composite systems and AGNs. Our study is based on a multi-wavelength analysis of a sample of 65 SWIRE-selected IR luminous source candidates classified as starbursts, AGNs or composite objects for which MAMBO observations were carried out. According to the postulated scenario, larger stellar mass and lower SFRs are expected in AGNs with respect to the starbursts. Our analysis suggests that the stellar masses are consistent or even lower in AGNs compared to those in starbursts and composite systems. In the FIR–mm wavelength range, the AGN contribution becomes negligible and the host emission dominates in all mm-detected sources. The derived FIR luminosities, and associated SFRs, are consistent in all three classes of objects. We thus do not observe the expected trends, but the three types of objects share similar properties. There is one major difference, that is the presence of a hot dust component in AGNs that increases their 24–$70\,\mu m$ observed fluxes by up to a factor of 10 compared to the MIR emission in starbursts.

Our estimates of the relative timescales of the AGN and starburst phases indicate that AGN activity is present during 37% of the starburst lifetime and that it dominates the MIR emission during $\sim$7% of it. These results suggest that AGN activity does not follow the starburst phase, but occurs simultaneously with the starburst phase. Furthermore, the comparison between the stellar masses suggests that the most luminous AGN phase precedes the moderately luminous AGN phase (see also Polletta 2008), and that the intense radiation field produced by the AGN does not immediately halt either the accretion or the star-formation activity.

References

Bolzonella, M., Miralles, J.-M., & Pelló, R. 2000, *A&A*, **363**, 476
Coppin, K. E. K., Swinbank, A. M., Neri, R., *et al.* 2008, *MNRAS*, **389**, 45
Di Matteo, T., Springel, V., & Hernquist, L. 2005, *Nature*, **433**, 604
Fiolet, N., Omont, A., Polletta, M. *et al.* 2009, *A&A*, **508**, 117
Kennicutt, Jr., R. C. 1998, *ARA&A*, **36**, 189
Lonsdale, C. J. *et al.* 2003, *PASP*, **115**, 897
Lonsdale, C. J., Polletta, M., Omont, A., *et al.* 2009, *ApJ*, **692**, 422
Polletta, M. 2008, *A&A*, **480**, L41
Polletta, M., Tajer, M., Maraschi, L., *et al.* 2007, *ApJ*, **663**, 81
Weedman, D. W. & Houck, J. R. 2009, *ApJ*, **698**, 1682

4

Star formation in galaxies hosting active galactic nuclei up to $z \sim 1$

John D. Silverman, zCOSMOS & XMM-COSMOS

4.1 Introduction

This contribution aims to address the fundamental question, effectively highlighting the overall theme of the workshop, as to what processes are important for eventually suppressing the growth of supermassive black holes (SMBHs) and how is this related to the evolution of star formation from $z \sim 1$ to the present. As illustrated in Figure 4.1, a global decline in mass accretion onto SMBHs and star formation rate density over the last 8 Gyr (Boyle and Terlevich 1998; Merloni 2004; Silverman *et al.* 2008b) is evident and may be driven by a mechanism such as feedback from AGN affecting the gas content of their hosts (Granato *et al.* 2004; Hopkins *et al.* 2008; Silk and Norman 2009). Such coupling may not only explain the local SMBH–bulge relations (see Shankar 2009 for an overview) but rectify the inconsistency between the observed distribution of high-mass galaxies and that predicted by semi-analytic models (Croton *et al.* 2006). Intriguingly, there is observational evidence for AGNs influencing their larger-scale environments that may lend support for the aforementioned feedback models. For example, radio jets are capable of impacting their intracluster medium, (Fabian *et al.* 2006) which may then in turn regulate further cluster cooling and inhibit star formation in the AGN host galaxy itself (Rafferty *et al.* 2008). Even at low power, radio-emitting outflows are capable of redistributing line-emitting gas in galactic nuclei (Whittle and Wilson 2004). Furthermore, QSO-driven winds are a common phenomenon, having kinetic energies quite capable of expelling gas, especially from the nuclear region. However, it has not been demonstrated, to date, that AGNs or the more luminous QSOs are responsible for quenching star formation on galactic scales.

Recent studies are providing evidence that apparently indicates an impact of AGNs upon their host galaxies. For example, X-ray selected surveys

AGN Feedback in Galaxy Formation, eds. V. Antonuccio-Delogu and J. Silk. Published by Cambridge University Press. © Cambridge University Press 2011.

J. D. Silverman et al.

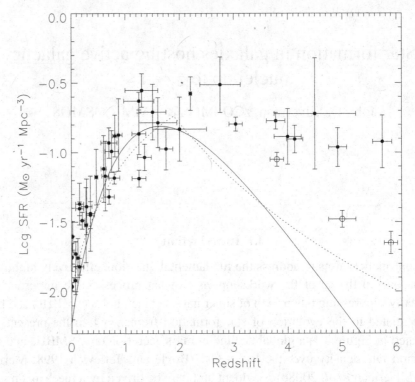

Figure 4.1 Global evolution of star formation rate density (data points) and mass accretion rate density onto SMBHs scaled up by a factor of 5000 for comparison (curves). (See Silverman *et al.* 2008a for details.)

(Nandra *et al.* 2007; Silverman *et al.* 2008b; Hickox *et al.* 2009; Schawinski *et al.* 2009) that utilize the obscured AGN population to enable a fairly clean view of the host galaxy demonstrate that many have rest-frame optical colors placing them in a region of the color–magnitude diagram usually populated with transitional galaxies (i.e., "green valley"; $U - V \sim 0.7$; see Figure 4.2a). This suggests that AGN feedback may contribute to the migration of galaxies from the "blue cloud" to the "red sequence". We note that the subject of the location of AGN hosts on the color–magnitude diagram may present an incomplete picture that will be addressed below (Silverman *et al.* 2009).

To further explore the role of SMBHs in galaxy evolution, we aim to determine whether AGNs are directly regulating the current rate of star formation, which may then depend on the accretion rate of the black hole itself. To do so, it is imperative to construct large samples of galaxies (> 10k) and compare ongoing star formation rates of galaxies hosting AGNs to those of the underlying galaxy population. The short duty cycles of accretion onto SMBHs Martini (2004) dictate the need for samples of such size. The spectroscopic nature for the sample will further

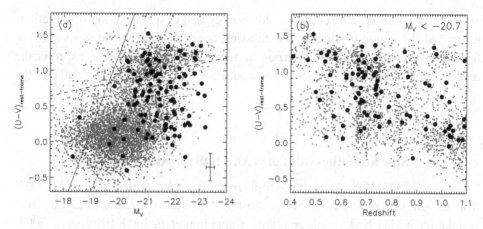

Figure 4.2 Host galaxy colors of AGNs in the Extended Chandra Deep Field – South survey: (a) rest-frame color versus absolute magnitude, (b) rest-frame color versus redshift. Galaxies with photometric redshifts from COMBO-17 (Wolf *et al.* 2004) are shown in gray while those hosting X-ray selected AGNs are marked by the larger black circles. (See Silverman *et al.* 2008b for further details.)

allow the disentanglement of environmental factors that are known to influence star formation. In addition, a multi-wavelength approach is also a necessity to characterize the intrinsic properties of the galaxy population (e.g. stellar mass). The advent of such surveys, starting with the Sloan Digital Sky Survey (SDSS), is enabling investigations of the relationship between galaxy and SMBH growth. Based on the enormity of the SDSS database, it has been clearly shown that galaxies hosting obscured AGNs have young stellar populations equivalent to late-type galaxies (Kauffmann *et al.* 2003, 2007) thus clearly establishing a direct relationship between SMBH accretion and star formation. On the contrary, differences between the stellar populations of AGN hosts and galaxies lacking AGN signatures in the SDSS have been reported (Martin *et al.* 2007), similar to the results based on X-ray selected AGNs, and are attributed to the suppression of star formation but may be due to selection (e.g. luminosity versus mass) as detailed below.

Recently, large-scale spectroscopic redshift surveys (e.g. DEEP2, COSMOS) have targeted the galaxy population out to higher redshifts ($z \sim 1.2$) thus beginning to probe the peak of star formation and AGN activity. A full multi-wavelength (UV-to-IR) approach is required to effectively characterize the galaxy population and its evolution. Equally important, the identification of galaxies at $z > 0.3$ that host obscured AGN demands an alternative selection technique (e.g. X-ray, IR) to optical emission-line diagnostics. Much progress has been made in recent years based on these deep multi-wavelength surveys to determine the host galaxy properties of AGNs up to $z \sim 1$ (e.g. Nandra *et al.* 2007; Gabor *et al.* 2009). In this review, we highlight our work using the COSMOS survey with specific attention to the zCOSMOS spectroscopic redshift survey and cospatial X-ray observations using XMM-Newton to identify the galaxies, based on their stellar mass and star formation rates, most likely to harbor an actively accreting SMBH. We refer the reader to the full publication (Silverman *et al.* 2009) that provides details on the methods and analysis techniques.

4.2 Star formation rates in zCOSMOS galaxies hosting AGN

We use the zCOSMOS 10k spectroscopic redshift catalog (Lilly *et al.* 2007, 2009) to investigate the properties of galaxies hosting AGNs and their relation to the parent population. XMM-Newton observations (Cappelluti *et al.* 2009; Brusa *et al.* 2007) of the full zCOSMOS sample enable us to identify 152 AGNs that include those with significant obscuration and of low optical luminosity. The derived properties such as host galaxy stellar mass, rest-frame color, and emission-line strength allow us to determine the prevalence of AGN activity as a function of these quantities. Specifically, we measure the SFR of galaxies using the [OII]λ3727 line luminosity. We account for the contribution from the underlying AGN component most likely arising from the narrow-line region by using the observed [OIII]λ5007 luminosity and the typical [OII]/[OIII] ratio found from previous studies of type 1 AGN (Ho 2005; Kim *et al.* 2006). The [OIII] line luminosity is measured directly from our spectra if present. For the subsample with [OIII] outside our observed spectral bandpass, we infer the [OIII] strength from the hard (2–10 keV) X-ray luminosity and the well-known correlation between these two quantities. Again, we refer the reader to Silverman *et al.* (2009) for full details regarding the method and subsequent results.

We find based on a stellar mass-selected sample of galaxies ($M_* > 4 \times 10^{10} M_\odot$) that significant levels of star formation are present in the hosts of X-ray selected AGNs (Figure 4.3). SFRs (1) range from $\sim$1 to 100 $M_\odot$ yr^{-1}, with an average SFR higher than that of galaxies with equivalent stellar mass, and (2) evolve with cosmic time in a manner that closely mirrors the overall galaxy population. Such evolution appears to be consistent with the low SFRs in AGNs ($z < 0.35$) from the SDSS. Therefore, we find no evidence for significantly reduced levels of star formation in the hosts of AGNs and conclude that massive galaxies with plentiful gas supplies are most conducive for AGN activity. This analysis effectively extends the clear association of AGN activity and star formation, seen in low-redshift studies with the SDSS (Kauffmann *et al.* 2003), up to $z \sim 1$. We note that similar findings are obtained here with additional spectral indicators, in particular $D_n(4000)$ and rest-frame optical colors $U - V$ (see the following section). Finally, we highlight that these results are consistent with the color evolution of AGN hosts, seen in the

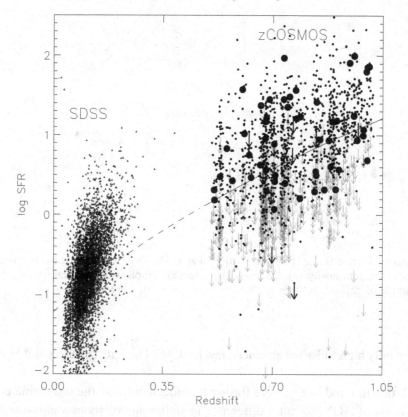

Figure 4.3 Star formation rate versus redshift for zCOSMOS galaxies ($0.5 < z < 1.0$) with those hosting AGNs marked by the large black circles. Upper limits are shown by the arrows. For comparison, type 2 AGNs from the SDSS are shown at $z < 0.35$ (small dots) (Kauffmann *et al.* 2003).

Chandra Deep Field – South survey (Silverman *et al.* 2008b), which follows that of the underlying galaxy population (see Figure 4.2b).

4.3 Further remarks on color–magnitude diagrams of AGN hosts

Much emphasis has been placed recently on the fact that AGN host galaxies have rest-frame optical colors between those of blue, star-forming galaxies and those of redder evolved galaxies (e.g. Nandra *et al.* 2007; Silverman *et al.* 2008b; Schawinski *et al.* 2009). Such observations have been thought to lend support for the role of AGNs in quenching star formation; although a deficiency of AGNs within the blue galaxy population appears to be in disagreement with the SFRs of AGN hosts in zCOSMOS galaxies presented above, and also with the fact that local ULIRGS

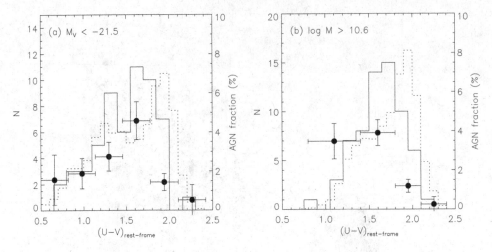

Figure 4.4 Rest-frame color distribution of zCOSMOS galaxies and those hosting
AGNs for luminosity (a) and mass (b) selected samples. Data points show the
fraction of galaxies hosting AGN with $L_{0.5-8\text{keV}} \sim 10^{43}$ erg s^{-1}.

have not only high SFRs but enhanced levels of AGN activity (Sanders and Mirabel
1996).

With this in mind, we venture further to understand why the rest-frame colors
of the hosts of AGN exhibit a difference in stellar age of about a gigayear from
that of star-forming galaxies. It is suspected that the discrepancy arises due to mass
selection since the hosts of type 2 AGNs in SDSS (Kauffmann *et al.* 2003) do not
exhibit such a difference with late-type galaxies of equivalent stellar mass. To check
this, we simply determined the fraction of galaxies hosting X-ray selected AGNs as
a function of rest-frame color for both a luminosity and mass selected sample. As
shown in Figure 4.4, the difference between the two selection methods is in the
fraction of blue ($U - V < 1.5$) galaxies hosting AGNs. The mass selection miti-
gates the inclusion of galaxies having low mass-to-light ratios that mainly pertains
to those forming stars. Since the incidence of AGN activity is known to rise with the
stellar mass of its host galaxy (See Figure 7 of Silverman *et al.* 2009), the decline in
AGN fraction from the "green valley" to the "blue cloud" seen in luminosity limited
samples is driven by the preponderance of low mass galaxies that are not likely
to harbor AGN of these X-ray luminosities. Simply put, the dependence of AGN
activity on stellar mass must be taken into consideration before making claims
regarding the color dependency of AGN activity in luminosity-selected samples.
In light of this check for consistency, we conclusively find that not only our SFRs
based on [OII] but also the rest-frame colors $U - V$ and spectral index $D_n(4000)$
all indicate that AGNs prefer to reside in galaxies with substantial levels of star

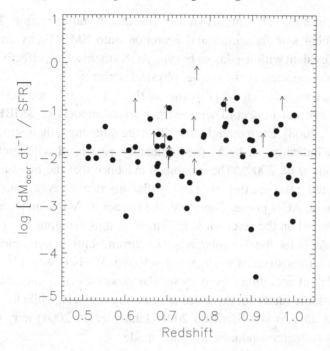

Figure 4.5 Ratio of mass accretion to SFR versus redshift. The horizontal dashed line marks the median ratio. Measurements are shown by a solid circle while lower limits are given by an arrow.

formation, in agreement with related studies at similar redshifts (Lehmer *et al.* 2008; Alonso-Herrero *et al.* 2009; Kiuchi *et al.* 2009).

4.4 Conclusion: co-evolution of SMBHs and their host galaxies

We have demonstrated that star formation with rates between ~ 1 and 100 $M_\odot$ yr^{-1} is present in the hosts of AGN up to $z \sim 1$. The question now is, how closely does star formation track the mass accretion rate onto these SMBHs? To answer this question, we have converted the X-ray luminosity to a mass accretion rate assuming a bolometric correction and accretion efficiency. In Figure 4.5, we plot the relative growth rate of these SMBHs to their host galaxies (dM_{accr} dt^{-1}/SFR). We find that a significant amount of dispersion is present, thus indicating that a direct relationship between star formation and black hole accretion does not occur on a case-by-case basis (see Figure 13a of Silverman *et al.* 2009). On average, a co-evolution scenario is clearly evident given the constancy of this ratio ($\sim 10^{-2}$) with redshift. Remarkably, this ratio is in very good agreement with that of low redshift type 2 AGNs in SDSS (Netzer 2009). The order-of-magnitude increase in this ratio, compared to the locally measured value of $M_{\mathrm{BH}}/M_{\mathrm{bulge}}$, is consistent with an AGN lifetime substantially shorter than that of star formation. This mutual decline in global star formation and accretion onto SMBHs, as introduced in Figure 4.1, is evident within galaxies hosting AGNs themselves, effectively shifting such a co-evolution scenario to smaller physical scales.

Overall, we conclude that the properties of these X-ray selected AGNs and their host galaxies are not in accord with merger-driven models of SMBH accretion (Hopkins *et al.* 2008). Even though their SFRs are quite high, their structural properties are not indicative of being predominantly associated with such disturbed systems (Gabor *et al.* 2009). There is much evidence that the hosts are massive, bulge-dominated galaxies, thus suggesting that any merger event must have happened prior to an AGN phase. Therefore, the impact of AGNs on their hosts may be minimal, based on the levels of star formation, thus bringing into question the efficiency of AGN feedback implemented in current semi-analytic models. Given the moderate luminosities of these X-ray selected AGNs ($L_X \sim 10^{43}$ erg s^{-1}), a "Seyfert mode" of accretion driven by secular processes (Hopkins and Hernquist 2006) may be more likely for this class of accreting SMBHs, while the more luminous QSOs (Canalizo and Stockton 2001; Jahnke *et al.* 2004) may provide the missing link to a merger-induced accretion mode.

References

Alonso-Herrero, A., Rieke, G., Colina, L. *et al.* (2009). The host galaxies and black holes of typical z 0.5–1.4 AGNs. *ApJ*, **677**, 127

Boyle, B. J. & Terlevich, R. J. (1998). The cosmological evolution of the QSO luminosity density and of the star formation rate. *MNRAS*, **293**, 49

Brusa, M., Zamorani, G., Comastri, A. *et al.* (2007). The XMM-Newton wide-field survey in the COSMOS field. III. Optical identification and multiwavelength properties of a large sample of X-ray-selected sources. *ApJS*, **172**, 353

Canalizo, G. & Stockton, A. (2001). Quasi-stellar objects, ultraluminous infrared galaxies, and mergers. *ApJ*, **555**, 719

Cappelluti, N., Brusa, M., Hasinger, G. *et al.* (2009). The XMM-Newton wide-field survey in the COSMOS field. The point-like X-ray source catalogue. *A&A*, **497**, 635

Croton, D. J., Springel, V., White, S. *et al.* (2006). The many lives of active galactic nuclei: cooling flows, black holes and the luminosities and colours of galaxies. *MNRAS*, **365**, 11

Fabian, A., Sanders, J. S., Taylor, G. B. *et al.* (2006). A very deep Chandra observation of the Perseus cluster: shocks, ripples and conduction. *MNRAS*, **366**, 417

Gabor, J. M., Impey, C. D., Jahnke, K. *et al.* (2009). Active galactic nucleus host galaxy morphologies in COSMOS. *ApJ*, **691**, 705

Granato, G. L., De Zotti, G., Silva, L., Bressan, A., Danese, L. (2004). A physical model for the coevolution of QSOs and their spheroidal hosts. *ApJ*, **600**, 580

Hickox, R., Jones, C., Forman, W. *et al.* (2009). Host galaxies, clustering, Eddington ratios, and evolution of radio, X-ray, and infrared-selected AGNs. *ApJ*, **696**, 891

Ho, L. (2005). [O II] emission in quasar host galaxies: evidence for a suppressed star formation efficiency. *ApJ*, **629**, 680

Hopkins, P. F. & Hernquist, L. (2006). Fueling low-level AGN activity through stochastic accretion of cold gas. *ApJS*, **166**, 1

Hopkins, P. F., Hernquist, L., Cox, T., Keres, D. (2008). A cosmological framework for the co-evolution of quasars, supermassive black holes, and elliptical galaxies. I. Galaxy mergers and quasar activity. *ApJS*, **175**, 356

Jahnke, K., Sanchez, S. F., Wisotzki, L. *et al.* (2004). Ultraviolet light from young stars in GEMS quasar host galaxies at $1.8 < z < 2.75$. *ApJ*, **614**, 568

Kauffmann, G., Heckman, T., Tremonti, C. *et al.* (2003). The host galaxies of active galactic nuclei. *MNRAS*, **346**, 1045

Kauffmann, G., Heckman, T., Budacary, T. *et al.* (2007). Ongoing formation of bulges and black holes in the local universe: new insights from GALEX. *MNRAS*, **173**, 357

Kim, M., Ho, L., Im, M. (2006). Constraints on the star formation rate in active galaxies. *ApJ*, **642**, 702

Kiuchi, G., Ohta, K., Akiyama, M. (2009). Co-evolution of supermassive black holes and host galaxies from $z \sim 1$ to $z = 0$. *ApJ*, **696**, 1051

Lehmer, B. D., Brandt, W. N., Alexander, D. M. *et al.* (2008). Tracing the mass-dependent star formation history of late-type galaxies using X-ray emission: results from the Chandra deep fields. *ApJ*, **681**, 1163

Lilly, S. J., Le Fèvre, O., Renzini, A. *et al.* (2007). zCOSMOS: a large VLT/VIMOS redshift survey covering $0 < z < 3$ in the COSMOS field. *ApJS*, **172**, 70

Lilly, S. J., Le Fèvre, O., Renzini, A. *et al.* (2009). The zCOSMOS 10k-bright spectroscopic sample. *ApJ* submitted

Martin, D. C., Wyder, T. K., Schiminovich, D. *et al.* (2007). The UV-optical galaxy color-magnitude diagram. III. Constraints on evolution from the blue to the red sequence. *ApJS*, **173**, 342

Martini, P. (2004). QSO lifetimes, in *Coevolution of black holes and galaxies* (Carnegie Observatories Astrophysics Series 1), ed. L. C. Ho, Cambridge: Cambridge University Press, p.169

Merloni, A. (2004). Tracing the cosmological assembly of stars and supermassive black holes in galaxies. *MNRAS*, **354**, 37

Nandra, K., Georgakakis, A., Willmer, C. N. *et al.* (2007). AEGIS. The color-magnitude relation for X-ray-selected active galactic nuclei. *ApJ*, **660**, 11

Netzer, H. (2009). Accretion and star formation rates in low redshift type-II active galactic nuclei, *MNRAS*, in press, arXiv:0907.3575

Rafferty, D. A., McNamara, B. R., Nulsen, P. E. J., Wise, M. W. (2008). The regulation of cooling and star formation in luminous galaxies by active galactic nucleus feedback and the cooling-time/entropy threshold for the onset of star formation. *ApJ*, **687**, 899

Sanders, D. B. & Mirabel, I. F. (1996). Luminous infrared galaxies. *ARA&A*, **34**, 749

Schawinski, K., Virani, S., Simmons, B. *et al.* (2009). Do moderate-luminosity active galactic nuclei suppress star formation?. *ApJ*, **692**, 19

Shankar, F. (2009). The demography of super-massive black holes: growing monsters at the heart of galaxies, *New Astronomy Reviews*, in press, arXiv:0907.5213

Silk, J. & Norman, C. (2009). Global star formation revisited. *ApJ*, **700**, 262

Silverman, J. D., Green, P. J., Barkhouse, W. *et al.* (2008a). The luminosity function of X-ray selected AGN: evolution of supermassive black holes. *ApJ*, **675**, 1025

Silverman, J. D., Mainieri, V., Lehmer, B. D. *et al.* (2008b). The evolution of AGN host galaxies: from blue to red and the influence of large-scale structures. *ApJ*, **679**, 118

Silverman, J. D., Lamareille, F., Maier, C. *et al.* (2009). Ongoing and co-evolving star formation in zCOSMOS galaxies hosting AGN. *ApJ*, **696**, 396

Whittle, M. & Wilson, A. (2004). Jet-gas interactions in Markarian 78. I. Morphology and kinematics. *ApJ*, **127**, 606

Wolf, C., Meisenheimer, K., Kleinheinrich, M. *et al.* (2004). A catalogue of the Chandra Deep Field South with multi-colour classification and photometric redshifts from COMBO-17. *A&A*, **421**, 913

Part II

Co-evolution of black holes and galaxies

5

The symbiosis between galaxies and SMBHs

G. L. Granato, M. Cook, A. Lapi & L. Silva

5.1 Introduction

A fundamental issue when modeling the evolution of galaxies in a cosmological context is that the majority of the processes driving baryonic evolution (such as star formation, various feedback mechanisms, accretion onto supermassive black holes (SMBHs)) operate or originate on scales well below the resolution of any feasible simulation in a cosmic box. Moreover, these processes are highly nonlinear, poorly understood from a physical point of view, and approximated by means of simplified, often phenomenological, and thus uncertain subgrid prescriptions. Unfortunately, yet unsurprisingly, a number of studies have clearly demonstrated that the results of these models are heavily affected by different choices for such prescriptions (e.g. Benson *et al.* 2003; Di Matteo *et al.* 2005), or for parameter values (e.g. Zavala *et al.* 2008). It is fair to say that first principles or *ab-initio* models do not exist.

5.2 Standard SAMs, their successes and their failures

Extensive comparisons between different scenarios and data are generally conducted by means of semi-analytic modeling (SAMs) for baryons, often grafted onto gravity-only simulations for the dark matter (DM) evolution. By the definition of SAMs, the general behavior of the system is outlined *a priori*, and then translated into a set of (somewhat) physically grounded analytical recipes – suitable for numerical computation over cosmological timescales – for the processes that are *thought* to be more relevant to galaxy formation and evolution.

Although SAMs should not be viewed as first-principles computations, they provide a convenient and powerful tool to test an assumed galaxy formation scenario

AGN Feedback in Galaxy Formation, eds. V. Antonuccio-Delogu and J. Silk. Published by Cambridge University Press. © Cambridge University Press 2011.

(i.e. the general behavior and the adopted recipes) against existing data, and to make predictions on future observations.

In general, SAMs (e.g. Cole *et al.* 2000; Hatton *et al.* 2003; Baugh *et al.* 2005; Cattaneo *et al.* 2005, 2006; Bower *et al.* 2006; Croton *et al.* 2006; Khochfar & Silk 2006; Monaco *et al.* 2007; Somerville *et al.* 2008), apart from relatively minor variations, are constructed around two main assumptions: (1) the initial outcome of gas cooling within DM halos is, at any cosmic epoch, the development of a rotationally supported disc (since Rees & Ostriker 1977; Silk 1977; White & Rees 1978); these discs usually undergo mild to moderate star formation activity, unless extreme choices for the scaling of star formation efficiency with galaxy properties are made; (2) the most natural driver of episodes of violent star formation at any redshift is the merging of these gas-rich discs, which in most models also constitutes the main channel for the formation of spheroids, and in particular of large ellipticals (since Cole 1991).

As a result of this disc-merger-driven framework, baryons tend to follow the hierarchical behavior of DM halos, and there is no inherent relationship between the morphology and the star formation history of galaxies. This is in sharp contrast with the basic observational fact that low-mass galaxies tend to be disc dominated, gas rich, blue, and actively star forming, whilst more massive galaxies tend to be red, gas poor, quiescent, and dominated by a spheroidal component mainly comprising old stars.

Due to these features, SAMs built around the two aforementioned assumptions – which from now on will be collectively referred to as "standard SAMs" – tend to be in discord with several observations (e.g. Somerville *et al.* 2008), manifested by the poor performance they had in anticipating observational breakthroughs that occurred more recently. For example, it is now well established that baryonic structures undergo the phenomenon referred to as "cosmic downsizing", whereby massive star-forming systems and associated SMBHs shine mostly at high redshift, while smaller objects display longer-lasting activity. Clearly, it is challenging to obtain this behavior from the scheme outlined above; indeed, no model did until relatively recently, and the present situation remains unclear. In the past few years, almost all semi-analytic teams introduced simple recipes of feedback from active galactic nuclei (AGNs) in their models, with the specific target of quenching star formation in high-mass galaxies at low redshift. This additional ingredient significantly improves the situation, but does not directly alleviate model difficulties in producing enough massive systems at high z. As a result, at least three state-of-the-art standard SAMs still do not correctly reproduce the downsizing trend in stellar mass, nor the *archeological downsizing* (Fontanot *et al.* 2009; see below for more details).

A further example of challenges to models comes from the modest evolution of the cosmic star formation activity above $z \sim 1$ (the so-called Madau plot), strikingly at variance with model predictions (e.g. Cole *et al.* 1994) generated before the advent of surveys effective in discovering dust-enshrouded star formation at high *z* (Madau *et al.* 1996, 1998). It is fair to note that a fraction, but not all, of the discrepancy was due to the then adopted standard CDM cosmology and a lower normalization for the fluctuation spectrum, resulting in significantly more rapid evolution at high redshift than the now favored ΛCDM.

In addition, even the latest and most refined SAMs are seriously challenged by the bright number counts and the high redshift peak of *z*-distribution for sub-mm galaxies. For instance, Baugh *et al.* (2005) showed that the only way to reproduce the statistic of sub-mm sources, usually considered the precursor of local ellipticals, in the context of their standard SAM, is to adopt an extremely top-heavy intial mass function (IMF) during galaxy-merger-induced starbursts. However, their model predicts masses of sub-mm sources likely too low by more than one order of magnitude (Swinbank *et al.* 2008), and still shows discrepancies with observed trends of α/Fe in local ellipticals (Nagashima *et al.* 2005).

Without doubt, the field of galaxy formation is *led* by observations. Indeed, physical processes have been continuously added to SAMs, or existing ones have been substantially revised by SAM developers in order to face serious mismatches between model outputs and new datasets. Besides many relatively minor but subtle details, major examples comprise a treatment of the growth of SMBHs in galaxy centers and of the ensuing energetic feedback from nuclear activity (Granato *et al.* 2004; Bower *et al.* 2006; Croton *et al.* 2006; Monaco *et al.* 2007; Somerville *et al.* 2008), the effects of "cold" versus "hot" accretion flows onto DM halos, as suggested by Dekel and Birnboim (2006) and implemented in a full SAM by Cattaneo *et al.* (2006; see also Somerville *et al.* 2008), or an extremely top-heavy flat IMF in merger-driven bursts (Baugh *et al.* 2005). These examples show that the complexity and degrees of freedom of standard SAMs have been steadily increased by modelers in order to improve the agreement with the data, but despite these efforts several points of tension still remain.

5.3 Possible solution from joint evolution of QSO and spheroids

A currently popular idea is that at least part of the solution could come from an ingredient that only very recently has started to be taken into account in some models, i.e. the mutual influence or feedback of star formation in galaxies and the development of SMBH-QSO at their centers. This influence is suggested by several empirical and also theoretical facts, such as the well-established local correlation between the mass of the central SMBH and several properties of the hosting spheroid (the luminosity, the mass, the velocity dispersion), the similarity of the cosmic development of SFR(z) and the luminosity of QSO per unit volume, or the fact that simulations of merging between galaxies drive flows of gas toward the central regions, creating an environment that is at least favorable to promote SMBH accretion. Having a relatively good determination of the almost constant ratio of stellar mass in spheroids and the mass of the hosted SMBH $\sim$1000, it is now interesting to compare the binding energy of spheroids with the energy released by the SMBH to accrete this mass. For a typical L^* galaxy, the latter turns out to be more than two orders of magnitude greater than the former, which means that a quite small coupling between the energy released by the AGN and the ISM is sufficient to have a significant effect. It is also interesting to compare the binding energy with the energy released by SNae during the assembly of the same spheroid, which is only about a factor 10 greater than the former.

The energy available is then large enough, though it is not clear if and how a (small) fraction of this energy can be transferred to the ISM. Several possibilities have been studied, in particular: radiation pressure, mostly on dust grains – a dusty medium is orders of magnitude more effective in absorbing momentum than a dust-free one; Compton or line radiative heating; and kinetic outflows from AGN, such as those observed in jets and broad absorption lines (likely generated by radiation pressure on resonant lines relatively close to the central engine). Another proposed possibility is that the AGN may act as a kind of catalyst to enhance the effectiveness of SNae feedback (Monaco 2004).

Only in the last few years have these effects begun to be explicitly considered in SAMs. This has been done along two quite distinct lines that should not be confused.

- Granato *et al.* (2004), Monaco and Fontanot (2005) and Menci *et al.* (2006) considered the feedback associated with the main phase of the BH growth, related to the bright high-z QSOs, as a way to sterilize massive high-z galaxies, which instead are little affected by SNae feedback, due to the depth of their potential well.
- More recently, and in the context of more standard SAMs, the feedback generated by lower redshift, lower accretion phases of AGN, in which almost all the accretion energy is used to halt cooling flows and avoid overproduction of local bright galaxies has been considered (e.g. Bower *et al.* 2006, Croton *et al.* 2006).

In general in this second set of works little attempt, or none at all, has been made to treat the build-up of SMBH and the physical nature of the feedback.

5.4 The ABC scenario

The first SAM in which a key role has been invoked for the reciprocal feedback between star formation (SF) and AGN activity is the "antihierarchical baryonic collapse" (ABC) proposed by Granato *et al.* (2004; see also Granato *et al.* 2001 for a more phenomenological treatment). This model, which is embedded in the ΛCDM hierarchical growth of DM halos and is focused on the formation of spheroids, adopts prescriptions to describe the baryonic physics that reverse the order with which spheroidal galaxies and high-z QSOs complete their formation, as indicated by the various evidence of down-sizing. This is obtained by a combination of two ingredients: (1) revised prescriptions for the SF in massive high-redshift galactic halos ($M_{vir} > 10^{12} M_{\odot}$), which allow SF rates as high as thousands of solar masses per year, as implied by observations of sub-mm galaxies, and (2) the inclusion of a treatment of the growth by accretion of a SMBH promoted by this huge SF activity (positive feedback between SF and accretion), which at some point becomes sufficiently powerful to clean the ISM and quench any further SF and accretion (negative QSO feedback). The ABC scenario predicts a well-defined evolutionary sequence leading to local ellipticals with dormant SMBH. The sequence begins with a high-redshift phase of huge, dust-enshrouded SF activity best detectable in the sub-mm spectral region, and lasting about 0.5 Gyr. This phase is ended by the strong feedback generated by the QSO phase, due to the growth of a SMBH, which is promoted by the huge SF activity during the previous SMG phase. This feedback sterilizes the system, which then evolves almost passively, thus the model predicts a sizeable population of massive and dead galaxies at high-z. The model leads (in one shot) to predictions in general agreement with many observations, which is rather disturbing for traditional SAMs.

For a description of the model and how it compares with observations of these populations of objects the reader is referred to Granato *et al.* (2001, 2004, 2006), Silva *et al.* (2005), and Lapi *et al.* (2006). More recently we have extended the computations to explicitly include also the formation and evolution of the disc component of galaxies (Cook *et al.* 2009).

Work in part supported by EC under CONTRACT MRTN-CT-2004-503929.

References

Baugh, C.M., *et al.* 2005, *MNRAS*, **356**, 1191
Benson, A.J., *et al.* 2003, *ApJ*, **599**, 38
Bower, R.G., *et al.* 2006, *MNRAS*, **370**, 645
Cattaneo, A., Blaizot, J., Devriendt, J., & Guiderdoni, B. 2005, *MNRAS*, **364**, 407
Cattaneo, A., *et al.* 2006, *MNRAS*, **370**, 1651
Cook, M. Lapi, A., & Granato, G.L. 2009, *MNRAS*, **397**, 534
Cole, S.M., Lacey, C.G., Baugh, C.M., & Frenk, C.S. 2000, *MNRAS*, **319**, 168
Cole, S.M., *et al.* 1994, *MNRAS*, **271**, 781
Cole, S.M. 1991, *ApJ*, **367**, 45
Croton, D.J., *et al.* 2006, *MNRAS*, **365**, 11
Di Matteo, T., Springel, V., & Hernquist, L. 2005, *Nature*, **433**, 604
Dekel, A., & Birnboim, Y. 2006, *MNRAS*, **368**, 2
Fontanot, F., *et al.* 2009, *MNRAS* (preprint arXiv:0901.1130)

Granato, G., *et al.* 2001, *MNRAS*, **324**, 757
Granato, G.L., *et al.* 2004, *ApJ*, **600**, 580 [G04]
Granato, G.L., *et al.* 2006, *MNRAS*, **368**, 72
Hatton, S. 2003, *MNRAS*, **343**, 75
Khochfar, S., & Silk, J. 2006, *ApJ*, **648**, L21
Lapi, A., *et al.* 2006, *ApJ*, **650**, 42
Madau, P., Pozzetti, L., & Dickinson, M. 1998, *ApJ*, **498**, 106
Madau, P., *et al.* 1996, *MNRAS*, **283**, 1388
Menci, N., Fontana, A., Giallongo, E., Grazian, A., & Salimbeni, S. 2006, *ApJ*, **647**, 753
Monaco, P. G. 2004, *MNRAS*, **352**, 181
Monaco, P. G., & Fontanot, F. 2005, *MNRAS*, **359**, 283
Monaco, P., Fontanot, F., & Taffoni, G. 2007, *MNRAS*, **375**, 1189
Nagashima, M., *et al.* 2005, *MNRAS*, **363**, L31
Rees, M.J., & Ostriker, J.P. 1977, *MNRAS*, **179**, 541
Silk, J.S. 1977, *ApJ*, **211**, 638
Silva, L., *et al.* 2005, *MNRAS*, **357**, 1295
Somerville, R.S., *et al.* 2008, *MNRAS*, **391**, 481
Swinbank, A.M., *et al.* 2008, *MNRAS*, **391**, 420
White, S.D.M., & Rees, M.J. 1978, *MNRAS*, **183**, 341
Zavala, J., Okamoto, T., & Frenk, C.S. 2008, *MNRAS*, **387**, 364

6

On the origin of halo assembly bias

A. Keselman

6.1 Introduction

Halo assembly bias, the dependence of dark-halo clustering on their formation history, is becoming increasingly important. The reason for this is that a better understanding of the formation of galaxies is needed in order to fully exploit new measurements, which are being developed with increasing precision.

According to the standard cosmological scenario, galaxies are formed in high-density regions consisting of virialized dark-matter particles. Such systems, termed dark-matter halos, form in a hierarchical and self-similar fashion, in which smaller objects form first and then continuously merge into ever larger objects. The merging process is not linear, in the sense that it doesn't arise from linear theory. Thus, it defines a time-dependent scale in which matter clustering becomes non-linear.

The theoretical framework describing the process of halo formation is the excursion set theory (Bond *et al.* 1991; Lacey & Cole 1993; Mo & White 1996). According to this theory, dark-matter density fluctuations, at a given scale, grow in the linear regime until they reach a critical value when they collapse. The collapse process is equivalent to a merging process of the smaller scales. When combined with the theory of random Gaussian fields, this framework explains the formation history of halos. A major result is that the history of a halo should not be correlated with the halo environment. This follows because density fluctuations are not correlated with larger scales. Another way of stating this is that halo history should not be correlated with the clustering of the halos themselves. And indeed, an assumption often made is that the clustering properties of halos depend on their mass alone. This has been confirmed by results of N-body simulations of intermediate resolution (Lemson & Kauffmann 1999; Percival *et al.* 2003).

AGN Feedback in Galaxy Formation, eds. V. Antonuccio-Delogu and J. Silk. Published by Cambridge University Press. © Cambridge University Press 2011.

The most popular quantification of halo assembly history is the halo formation age, or assembly age, where the age is defined as the time from formation until today, and the formation time is defined as that time when the halo mass was half of its current mass. Using this definition and the Millennium Simulation (Springel *et al.* 2005), Gao *et al.* (2005) show that the clustering of halos depends strongly on their age. They have found that the "oldest" 10% of the halos with mass $10^{11}h^{-1}M_\odot$ are more than five times more correlated than the "youngest" 10% of halos of the same mass. This assembly bias has been confirmed by Harker *et al.* (2006) using marked correlation functions on the same simulation, and by Wechsler *et al.* (2006) and Jing *et al.* (2007) using independent simulations. Wetzel *et al.* (2007) also found dependence of clustering on halo history, but only when using a different definition for the assembly redshift.

We still lack a completely satisfactory explanation for the origin of assembly bias. For Gaussian initial conditions, simple arguments based on the spherical collapse model applied to narrow and broad initial density peaks that would collapse to halos of the same mass at the present time do predict an assembly bias, but with younger halos being more clustered than older ones. This trend of the bias is opposite to what is seen in simulations. Tidal stripping has also been invoked as a possible mechanism (e.g. Diemand *et al.* 2007). Because of mass stripping by the tidal gravitational field of the large mass concentration, nearby halos would have been of higher mass in a different environment. Therefore, these halos would have earlier formation times and would be more biased than halos of the same mass in the field. Avila-Reese *et al.* (2005) suggested tidal stripping as a mechanism responsible for generation of assembly bias in the high-density regions, whereas in low-density regions the cosmological initial conditions play a more important role. Maulbetsch *et al.* (2007), Wang *et al.* (2006) and Desjacques (2008) suggest that the halo mass-accretion is less efficient in denser regions due to large-scale tidal fields. However, the extent of this effect is difficult to assess.

Keselman and Nusser (2007, hereafter KN07) have shown that the bias can, at least partially, arise in the quasi-linear evolution (i.e. over scales where the flow is still laminar). In order to eliminate highly non-linear effects such as tidal stripping, they used approximate methods based on the Zel'dovich approximation (Zel'dovich 1970), where particles move on straight lines independently of the motion of other particles. The Zel'dovich approximation is an analytic solution to planar cosmological perturbations up to the stage where multi-streaming appears. For three-dimensional perturbations, the approximation is a reasonable description of quasi-linear dynamics away from multi-streaming regions (e.g. Nusser *et al.* 1991). In order to extend the applicability of this approximation beyond multi-streaming, the following scheme was adopted. Particles initially move in straight lines according to Zel'dovich, but they merge together when they come within a critical distance of each other. This merging (sticking) produces an object with mass and linear momentum equal to the total of its components. The critical distance is taken to depend on time like a diffusion length, as inspired by the adhesion approximation. This is known as the punctuated Zel'dovich (PZ) approximation (Fontana *et al.* 1995). This approximation is ideal for the purposes of this study as it does not incorporate highly non-linear effects and also it is fast and easy to implement. Further, it readily provides merging trees for individual objects.

6.2 Measuring assembly bias in the quasi-linear regime

KN07 have run 60 PZ simulations, with a total particle number comparable to the Millennium Simulation. The initial conditions correspond to a random Gaussian realization of the cold dark matter scenario in a flat universe without a cosmological constant. The dependence on the background density parameters comes through the initial power spectrum, but the dynamics is nearly independent of these parameters when the linear growth factor is used as the time variable (Nusser & Colberg 1998). Therefore, apart from the effect of the initial power spectrum, the final result should be independent of the cosmological background.

For purposes of history tracking, each halo was assigned a unique ID. When halos merge, the newly formed halo inherits the ID of the most massive progenitor.

At the final time, the average number of objects per simulation was 7×10^5 with 10^5 being more massive than $4.3 \times 10^{11} h^{-1} M_\odot$, which is the minimal halo mass considered for the study of assembly bias. The conclusions were based on correlations on scales larger than $2h^{-1}$ Mpc, because of resolution effects. Note also that objects ("halos") in a PZ simulation are point-like and are identified using different criteria than halos in full N-body simulations. Only halos containing more than 100 particles $(4.3 \times 10^{11} h^{-1} M_\odot)$ at the final time $(z = 0)$ were considered. These halos were used to plot the different correlation functions, as functions of different formation age bins.

The dependence of the correlation function on the formation time is clear for all mass ranges. The bias persists even between the youngest 30% and oldest 30% of halos. The difference between the correlation functions of old and young halos was used to quantify the assembly bias at various separations. For masses $\lesssim 2-3 \times 10^{12} M_\odot$, the bias is about 1.7 and is similar for all separations considered here. The error-bars are large at separations $>10h^{-1}$ Mpc and an increase in the bias (as claimed by Gao *et al.* 2005) cannot be detected. The bias weakens with increasing halo mass, but remains statistically significant only for the 10% old/young halos, at separations $\lesssim 8h^{-1}$ Mpc. The figure shows that the mass scale $2-3 \times 10^{12} M_\odot$ marks a mass threshold above which assembly bias weakens, for all separations. This threshold is close to the non-linear mass scale M_*, defined as the mass scale over which the rms of density fluctuations is 1.69. For the initial conditions used, $M_* \approx 5 \times 10^{12} h^{-1} M_\odot$.

Assembly bias may be caused by the different environments of old and young halos. These may be found by cross-correlating the bias with several statistical measurements of the environment. The most relevant measure found was the "dimensionality" of the density field in regions near halo particles at the initial time. This parameter is an indicator of the geometry of the structure developing at later times in those regions. At the centers of spherical, cylindrical, and planar perturbations the dimensionality obtains the values $\sqrt{3}$, $\sqrt{2}$, and 1, respectively. The mean value of the dimensionality as a function of distance from particles making up young and old halos was plotted. The results show clearly that young halos have a higher average dimensionality than old ones.

6.3 Conclusions

Assembly bias of halos persists even in a simplified description of gravitational dynamics such as the punctuated Zel'dovich (PZ) approximation. The PZ approximation prevents the coasting away of particles in multi-streaming regions by coalescing objects that have come within a critical distance of each other. The PZ is fast, simple to implement, and readily provides object merging trees. This allowed the study of assembly bias in a large number of simulations (60 simulations, each of 512^3 particles in a $(128 h^{-1} \text{Mpc})^3$ cubic box). The magnitude of the bias is comparable to that found in full N-body simulations. This implies that highly non-linear effects, such as mass loss from halos in the vicinity of larger mass concentrations, may not be the dominant mechanism for production of assembly bias.

There is a strong correlation between halo ages and the dimensionality of the nearby initial configuration. Young halos tend to form in regions of higher initial dimensionality than old halos. This is explained by the dependence of collapse time on dimensionality – a spherical perturbation collapses slower than a planar perturbation with the same initial density (Bertschinger & Jain 1994).

References

Avila-Reese V., Colín P., Gottlöber S., Firmani C., Maulbetsch C. (2005). *ApJ*, **634**, 51
Bertschinger E., Jain B. (1994). *ApJ*, **431**, 486
Bond J. R., Cole S., Efstathiou G., Kaiser N. (1991). *ApJ*, **379**, 440
Desjacques V. (2008). *MNRAS*, **388**, 638
Diemand J., Kuhlen M., Madau P. (2007). *ApJ*, **667**, 859
Fontana L., Milelli M., Murante G., Provenzale A. (1995). *Il Nuovo Cimento C*, **18**, 530
Gao L., Springel V., White S. D. M. (2005). *MNRAS*, **363**, 66
Harker G., Cole S., Helly J., Frenk, C., Jenkins A. (2006). *MNRAS*, **367**, 1039

Jing Y. P., Suto Y. & Mo H. J. (2007). *ApJ*, **657**, 664
Keselman J. A., Nusser A. (2007). *MNRAS*, **382**, 1853
Lacey C., Cole S. (1993). *MNRAS*, **262**, 627
Lemson G., Kauffmann G. (1999). *MNRAS*, **302**, 111
Maulbetsch C., Avila-Reese V., Colín P., *et al.* (2007). *ApJ*, **654**, 53
Mo H. J., White S. D. M. (1996). *MNRAS*, **282**, 347
Nusser A., Dekel A., Bertschinger E., Blumenthal G. R. (1991). *ApJ*, **379**, 6
Nusser A., Colberg J. M. (1998). *MNRAS*, **294**, 457
Percival W. J., Scott D., Peacock J. A., Dunlop J. S. (2003). *MNRAS*, **338**, 31L
Springel V. *et al.* (2005). *Nature*, **435**, 629
Wang H. Y., Mo H. J., Jing Y. P. (2007). *MNRAS*, **375**, 633
Wechsler R. H., Zentner A. R., Bullock J. S., Kravtsov A. V., Allgood B. (2006). *ApJ*, **652**, 71
Wetzel A. R., Cohn J. D., White M., Holz D. E., Warren M. S. (2007). *ApJ*, **656**, 139
Zel'dovich Ya. B. (1970). *A&A*, **5**, 84

7

AGN, downsizing and galaxy bimodality

M. J. Stringer, A. J. Benson, K. Bundy & R. S. Ellis

7.1 Introduction

Only by incorporating various forms of feedback can theories of galaxy formation reproduce the present-day luminosity function of galaxies. It has also been argued that such feedback processes might explain the counterintuitive behaviour of 'downsizing' witnessed since redshifts $z \simeq 1 - 2$. To examine this question, observations spanning $0.4 < z < 1.4$ from the DEEP2/Palomar survey (Bundy *et al.* 2006) are compared with a suite of equivalent mock observations derived from the Millennium Simulation, populated with galaxies using the GALFORM code (Bower *et al.* 2006).

7.2 Hierarchical assembly

The mock galaxy samples are generated from the population of dark matter halos in the Millennium Simulation (Springel *et al.* 2005). This simulation consists of approximately 10 billion dark matter particles each of mass $8.6 \times 10^8 h^{-1} M_\odot$ evolving in a cubic volume of side $500 h^{-1}$ Mpc, assuming a ΛCDM cosmology.[1]

Dark matter halo merger trees are found from this 4-volume using the methods described by Harker *et al.* (2006). The lowest mass halos contained in these trees, of which there are about 20 million, consist of 20 particles corresponding to a total mass of $5 \times 10^9 h^{-1} M_\odot$. Such halos could contain at most $9 \times 10^8 h^{-1} M_\odot$ of baryonic material, which is well below the lower limit of the stellar mass functions to be considered in this work. Therefore we do not expect the resolution of the Millennium Simulation to affect our results.

---

[1] Specifically, a flat universe with $\Omega_b = 0.045$, $\Omega_M = 0.25$, $H_0 = 73$ km s^{-1}Mpc^{-1} and $\sigma_8 = 0.9$.

AGN Feedback in Galaxy Formation, eds. V. Antonuccio-Delogu and J. Silk. Published by Cambridge University Press. © Cambridge University Press 2011.

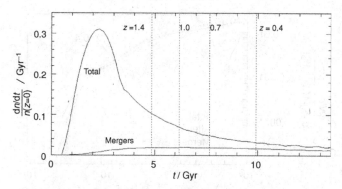

Figure 7.1 The formation rate of galaxies within the Millennium Simulation as a fraction of their present-day number. The upper line shows the total formation rate of galaxies with $M_\star > 10^9 M_\odot$; the lower line shows the rate at which pairs of these galaxies merge with each other. (Mergers involving smaller systems are not included.) The area between the curves will therefore be equal to one. Redshifts relevant to the DEEP2/Palomar samples are indicated with vertical dotted lines.

The assembly of dark matter halos in ΛCDM is often described as 'hierarchical'. This is appropriate in that some galaxies from one generation will merge to create the next. The importance of this contribution is illustrated in Figure 7.1, which demonstrates that the merger rate is expected to be quite small; less than 2% of galaxies with $M_\star > 10^9 M_\odot$ merge with each other every Gyr. At early times, minor mergers and the formation of new stars therefore create massive galaxies much faster than they can be destroyed by major mergers.

As the universe evolves, the creation rate of new galaxies diminishes, leading to a near zero net growth in numbers (the difference between the two curves in Figure 7.1). The main growth is now due to *differential* mass assembly, a trend that is illustrated in Figure 7.2. This shows the number of galaxies that would have formed by a given redshift through accretion alone (equivalent to the 'total' component in Figure 7.1) alongside the number that remain after all mergers (since $t = 0$) have taken place. The latter is the true, final number.

The number density of galaxies in a given mass range will rise or fall depending on whether more galaxies arrive or leave that range due to an increase in their stellar masses. At intermediate masses, the population is almost unaffected by mergers; the creation and destruction rates are approximately equal. At high mass, merging has a more significant effect because of the high ratio of potential progenitors to existing galaxies (indicated by a more steeply declining mass function at these masses). A particular deduction is that very massive galaxies ($M_\star > 3 \times 10^{11} M_\odot$) would be almost non-existent without merging.

The preceding discussion is primarily based upon the formation and interaction of dark matter halos in the simulation. However, in order to construct Figure 7.2 we

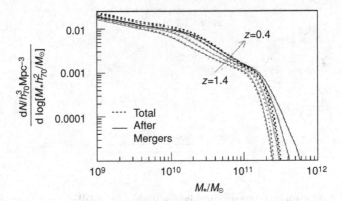

Figure 7.2 The evolving comoving number density of galaxies seen in the Millennium Simulation. Lines from top to bottom correspond to the redshift limits ($z = 0.4, 0.7, 1.0$ and 1.4) adopted in the DEEP2/Palomar survey. The total number of systems formed by a given redshift is shown as a dotted line whereas the associated solid line shows the true number, including mergers.

did have to consider the stellar content of the galaxies as well as their host halos. We now turn to discuss how stellar populations are introduced.

7.3 Modelling AGN feedback in galaxies

Halos in the merger trees introduced earlier are populated with galaxies using the GALFORM semi-analytic model (Cole *et al.* 2000). AGN heating has been introduced into this model (Bower *et al.* 2006), using the following simple treatment. A fraction $F_{BH} = 0.5\%$ of the disc gas is accreted on to the central black hole when there is either a merger or the disc's self-gravity exceeds the critical limit for stability:[2]

$$\sqrt{\frac{GM_{disc}}{r_{disc}}} > 0.8 V_{max}. \tag{7.1}$$

The surrounding hot halo is assumed to be hydrostatic if the free fall time is significantly shorter than the cooling time:

$$t_{cool} > \alpha_{cool} t_{ff} \qquad (\alpha_{cool} = 0.58). \tag{7.2}$$

If this condition is met, further cooling of gas is prevented when the Eddington luminosity of the super/massive black hole residing at the centre of the galaxy

---

[2] This criterion follows the work of Efstathiou *et al.* (1982), though the particular constant, 0.8, and the value of F_{BH}, were found by requiring the model to match the Magorrian relation between bulge mass and black hole mass, $M_{BH} \sim M_{bulge}^{1.12}$ (Haring and Rix 2004).

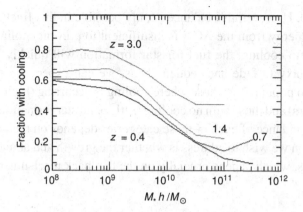

Figure 7.3 The fraction of galaxies with non-zero mass accretion due to cooling, plotted as a function of stellar mass for four redshifts, as labelled. The bottom three span the range of the DEEP2/Palomar samples. Significant cooling with subsequent star formation occurs for all masses at high redshift, declining to be almost completely suppressed for $M_* > 10^{11} M_\odot$ by $z \simeq 0.4$. This demonstrates how the continued suppression of star formation due to radio mode AGN feedback can reconcile downsizing in the context of hierarchical models.

greatly exceeds the cooling luminosity

$$L_{\text{Edd}} > \epsilon_{\text{SMBH}}^{-1} L_{\text{cool}} \qquad (\epsilon_{\text{SMBH}} = 0.04). \qquad (7.3)$$

The two free parameters, α_{cool} and ϵ_{SMBH}, were chosen (Bower *et al.* 2006) by constraining the model to match local luminosity function data. This strongly suppresses the formation of the most massive galaxies and imprints a near-exponential cut-off in the abundance of the brightest galaxies.

AGN heating is expected to primarily suppress the formation of the most massive galaxies. As a consequence of hierarchical growth, such systems will have considerable spheroidal components and, hence, massive central black holes. A study of AGN feedback implemented in semi-analytic models has been made (Croton *et al.* 2006), also using the Millennium Simulation, which produced qualitatively similar results.

Figure 7.3 illustrates the effect of AGN feedback in a simple way by showing the fraction of galaxies in the model with active cooling (i.e. those whose cooling has *not* been shut down by AGN heating). Cooling is largely unaffected in low-mass systems but is completely suppressed in the majority of massive systems. This illustrates how hierarchical structure formation can be made consistent with the observational phenomenon of downsizing.

In order for a galaxy to form a significant number of stars, there must be a supply of cooling gas from the surrounding halo to counteract the reheating of gas by the energy released. However, this balance is only possible in that fraction of galaxies for which the energy from the AGN is insufficient to prevent cooling of halo gas. In the case of no cooling, the fuel for star formation will quickly be exhausted, causing the galaxy to fade and redden. Thus, we can expect the galaxies to be divided into two populations: those where cooling is occurring (high star formation rate, blue colours) and those with no cooling (little or no star formation, red colours). The relative abundance of each of these categories depends on the galaxies' mass. The key question we wish to address is whether the growth and abundance of these two populations, which depend crucially on the feedback mechanism, match those observed.

7.4 Colour bimodality

We begin our comparison between the DEEP2/Palomar observations and the GAL-FORM model with an analysis of the colour distribution. Figure 7.4(left) plots, in three redshift intervals, the rest-frame $(U - B)$ colour–magnitude diagram of galaxies drawn from one set of mock lightcones constructed to match the properties of the DEEP2/Palomar dataset.

This re-enforces the previous claims (Bower *et al.* 2006; Croton *et al.* 2006) that the incorporation of radio-mode AGN feedback helps establish a distinct red sequence and prevents star formation in the brightest galaxies, as observed. However, the mock populations do not quite sit either side of the line that success-fully divided the observed populations.[3] A more profound discrepancy between the observed and modelled colours is shown in Figure 7.4(right), which plots the $(U - B)$ distribution in the same three redshift intervals. Results from the lightcones are illustrated by shaded curves while data points signify the observed distributions from the DEEP2/Palomar dataset. Beginning with the total distri-bution, we find that model galaxies trace a narrower range in $(U - B)$ colour than observed galaxies. Even after including photometric uncertainties of 0.1 mag, the distribution of colours is simply too narrow, particularly at low red-shift.

7.5 Understanding mass errors and cosmic variance

The stellar mass functions predicted by the model are plotted in Figure 7.5, showing good agreement with the observed total mass function *when the stel-lar mass uncertainties are convolved with the model results* (also emphasised in

---

[3] This line has a slope given by van Dokkum *et al.* (2006) with a vertical offset chosen empirically to divide the red sequence from the blue cloud in the observed DEEP2 distribution (Bundy *et al.* 2006; Willmer *et al.* 2006).

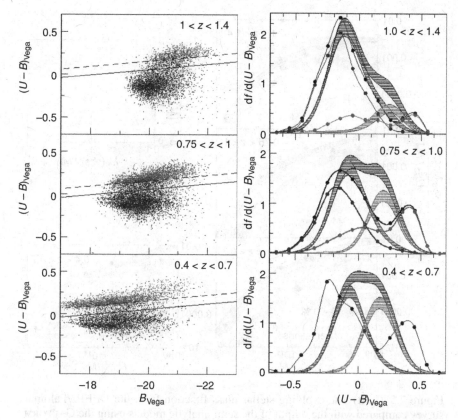

Figure 7.4 The predicted colour distribution of galaxies from one of the 20 light-cone samples. The left panel illustrates the distinction between star-forming and red-sequence galaxies. The solid diagonal line indicates the division made in the *observational* sample using rest-frame $U - B$ colours. The dashed line shows the division in the *model* sample, thereby producing Figure 7.5. The right panel shows the rest-frame $U - B$ colour distribution. Black points show observational values and shaded areas correspond to the total range of values found across 20 light-cones. For the latter, an observational error of 0.1 mag (1 s.d.) has been included. Both datasets are connected with smooth curves for visual clarity. The distinction between the two populations is highlighted by applying a cut on the star formation rate at $\dot{M}_\star = 0.2\ M_\odot\ \mathrm{Gyr}^{-1}$ (as in Bundy *et al.* (2006)). This criterion serves to illustrate the strong link between star formation rate and colour, but does not provide a precise division.

Kitsbichler and White (2007)). This has the effect of increasing the high-mass end of the predicted mass function, as can be seen by comparing the left and right panels of Figure 7.5. Fewer low-mass galaxies are observed than are predicted at all redshifts, despite the extremely efficient conversion of supernova energy to ejected gas in this model (Bower *et al.* 2006).

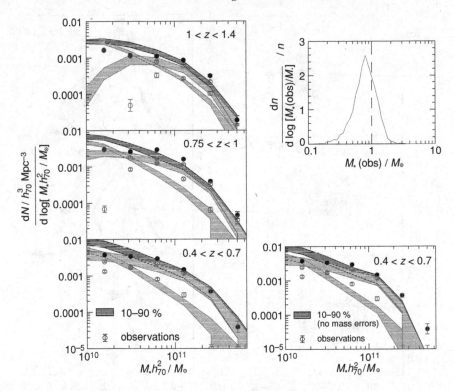

Figure 7.5 (Left) The evolving stellar mass function from the DEEP2/Palomar survey compared with the output of the semi-analytic models using the GALFORM radio-mode feedback. Included are the effects of cosmic variance calculated from 20 lightcones, reduced in size so that the four solid angles in each set correspond to those of the four DEEP2/Palomar survey fields (Stringer *et al.* 2009). Points are observational values from Bundy *et al.* (2006). The three sets of points and shaded regions refer to the total mass function (top) quiescent (lower) and star-forming (bottom) components, divided as shown in Figure 7.4. According to Bundy *et al.* (2006), the mass completeness limits from the K_S-band alone are $\log[M_\star/M_\odot] = 10.1$, 10.2 and 10.4 for the three respective redshift intervals. Note that, in this left panel, stellar masses include randomly generated errors to emulate those found in the observational technique, illustrated (right), upper panel. This shows stellar masses of a representative 15 000 galaxies, derived from the simulated photometry, plotted as a fraction of the generated stellar mass. The IMF of Chabrier (2003) was used in the model and in the analysis used to make the mass estimates. The importance of allowing for such errors can be appreciated by comparison with the lower panel.

Though disagreement between the predicted and observed total numbers appears to be minor, this thorough analysis of cosmic variance shows that there is still *significant inconsistency* with the data at many stellar mass intervals.

The substantial uncertainties from cosmic variance, which dominate the error budget, mean that much larger data sets will be required for more detailed studies capable of quantifying the importance of such effects as merging, internal growth due to SF, and transformations on the mass functions of different types of galaxies. Such work will represent an important step forward in understanding the physical nature of these mechanisms.

Acknowledgements

We would like to thank Carlton Baugh, Richard Bower, Shaun Cole, Carlos Frenk, John Helly, Cedric Lacey and Rowena Malbon for allowing us to use the GALFORM semi-analytic model of galaxy formation (www.galform.org).

The Millennium Simulation used in this paper was carried out by the Virgo Supercomputing Consortium at the Computing Centre of the Max-Planck Society in Garching. The databases and the web application providing online access to them were constructed as part of the activities of the German Astrophysical Virtual Observatory (www.g-vo.org/Millennium).

MJS acknowledges support from the Warden and Fellows of New College, Oxford, the hospitality of the CTCP at Caltech and of the KITP, Santa Barbara. AJB acknowledges support from the Gordon & Betty Moore Foundation. RSE acknowledges financial support from the Royal Society.

References

Bower, R. G., Benson, A. J., Malbon, R., *et al.* 2006, *MNRAS*, **370**, 645
Bundy, K., *et al.* 2006, *ApJ*, **651**, 120
Chabrier, G. 2003, *PASP*, **115**, 763
Cole, S., Lacey, C., Baugh, C., & Frenk, C. 2000, *MNRAS*, **319**, 168
Croton, D. J., *et al.* 2006, *MNRAS*, **365**, 11
Efstathiou, G., Lake, G., & Negroponte, J. 1982, *MNRAS*, **199**, 1069
Harker, G., Cole, S., Helly, J., Frenk, C. S., & Jenkins, A. 2006, *MNRAS*, **367**, 1039
Häring, N., & Rix, H.-W. 2004, *ApJ*, **604**, L89
Kitzbichler, M. G., & White, S. D. M. 2007, *MNRAS*, **376**, 2
Springel, V., White, S. D. M., & Jenkins, A. 2005, *Nature*, **435**, 629
Stringer, M. J., Benson, A. J., Bundy, K., Ellis, R. S., & Quetin, E. L. 2009, *MNRAS*, **393**, 1127
van Dokkum, P. G., *et al.* 2006, *ApJ*, **638**, L59
Willmer, C. N. A., *et al.* 2006, *ApJ*, **647**, 853

Part III

Outflows and radio galaxies

8

Interaction and gas outflows in radio-loud AGN: disruptive and constructive effects of radio jets

R. Morganti

8.1 Why radio-loud AGN?

In recent years, active galactic nuclei (AGN) have become more popular among a wider community. The possibility of using them to produce feedback effects that would help solve some of the questions connected to the hierarchical scenario of galaxy formation and evolution has made them particularly popular among theorists. Feedback effects associated with AGN-induced outflows are now routinely incorporated in models of galaxy evolution. Indeed, gas outflows may have a wide range of effects. For example, clearing up the circumnuclear regions and halting the growth of the supermassive black holes (see e.g. Silk & Rees 1998), as well as providing the mechanism that can cause the correlations found between the mass of the central super-massive black hole and the properties of the host galaxies. Outflows can also prevent the formation of too many massive galaxies in the early universe and can inject energy and metals into the interstellar and intergalactic medium. AGN-driven outflows could be a major source of feedback in the overall galaxy formation process. However, *the characteristics of such feedback are poorly constrained and the exact relevance of gaseous outflows in galaxy evolution still needs to be evaluated.*

AGN-driven outflows can have different origins. Here, I will concentrate on the role that the *radio-loud phase of nuclear activity* (and the presence of radio plasma jets) can play in the evolution of a galaxy. Radio-loud AGN are preferentially hosted by massive early-type galaxies. The fraction of these galaxies that are radio loud increases with mass: for the highest masses, the portion of galaxies that are radio sources is $\sim 25\%$. Considering that radio-loud AGN live for only 10^7–10^8 yr, the radio source activity must be constantly re-triggered

AGN Feedback in Galaxy Formation, eds. V. Antonuccio-Delogu and J. Silk. Published by Cambridge University Press. © Cambridge University Press 2011.

(Best *et al.* 2005 and references therein). Indeed, many examples are known of objects where signatures of recurrent radio activity are observed. They include famous cases, such as Centaurus A and 3C 236, and less famous cases associated with recently restarted radio sources (see e.g. Stanghellini *et al.* 2005). Thus, radio activity can be common in the life of *most* (if not all) early-type galaxies and may, therefore, be relevant in their evolution.

Before proceeding, it is also worth noting that, although the radio-loud sources considered in this paper are at low redshift and hosted by early-type galaxies, it does not mean that feedback effects are not relevant. Recent detailed studies of the ISM in nearby early-type galaxies have shown how these systems can be complex and that *the assumption that these systems are 'red and dead' and without an interesting and rich ISM does not hold for many of them.* Neutral hydrogen, molecular and ionised gas (see Morganti *et al.* 2006, Combes *et al.* 2007 and Sarzi *et al.* 2006, respectively) are often found (sometimes in large quantities) in these galaxies. This is also the case around and in the centre of, at least some, radio sources. The possible connection between the presence of gas and the presence (and type) of the radio source is also a matter of investigation (see e.g. Emonts *et al.* 2006). Relevant in the context of feedback and outflows is the fact that compact/young radio galaxies are more often detected in H I (both in emission and in absorption) and they show more extreme kinematics of the ionised gas (see e.g. Holt *et al.* 2008 and references therein) compared to extended/classical radio sources. Here, I will concentrate on two aspects in which radio activity could be important. The first is exploring whether the relativistic plasma jet associated with radio-loud galaxies could provide an effective way to produce gas outflows with characteristics that can be relevant in the evolution of the host galaxy. The second is to investigate whether they can provide a mechanism for the triggering of star formation.

The origin of nuclear activity is often explained as connected to merger and/or interaction processes. These processes can bring gas into the system, but they can also trigger radial motions that could lead to increased fuel rate. In about 30% of powerful radio galaxies, we know, from the analysis of the stellar population, that a gas-rich merger must have occurred in the recent history of the host galaxy. Although these are the most extreme cases, one can expect that, at least in the initial phase of activity, the supermassive black hole will often be surrounded by a rich gaseous medium (indeed observed in the case of young radio sources as mentioned above). The relativistic jets associated with the radio-loud phase of activity can provide a mechanism to directly couple the AGN to its environment and produce gaseous outflows. These outflows can therefore also be relevant for the orientation-independent obscuration, again providing one of the mechanisms (together with radiation pressure and starburst winds) to expel the gas that obscures the AGN in the initial phase If this is the case, the outflows would be particularly important for young AGN. Conversely, the ISM may also have an influence on the evolution of the radio jet: the interaction may cause a (temporary) disruption of the flow. Finally, jet-induced star formation has been suggested to be important for high-z objects and to be one of the possible causes of the alignment effect. This effect is not so commonly seen at low-z although there are a few cases known that have been studied in detail. In summary, the main aim of the projects described below is to understand what are the main sources of feedback and outflows in nearby objects and extrapolate this to the high-z universe, where usually the objects cannot be studied in such detail.

8.2 The nuclear regions probed by the H I and ionised gas

As mentioned above, gas is an important ingredient in the regions surrounding an active nucleus. Thus, we have used the gas (21 cm H I and ionised gas) to trace the effect of the radio plasma passing into the ISM of a radio source. Here we will summarise some of the best studied cases and the most relevant results.

8.2.1 Fast HI outflows: our best studied case – IC 5063

One of our best studied objects is the radio-loud Seyfert 2 galaxy IC 5063. Recent radio data (see Morganti *et al.* 2007 and references therein) have confirmed the triple structure of the source with a central, unresolved flat-spectrum core and two resolved radio lobes with steep spectral index (see Figure 8.1 left). This implies that the previously detected fast outflow of neutral gas is occurring off-nucleus, near the (brighter) radio lobe, i.e. about 0.5 kpc from the core. The ionised gas shows highly complex kinematics in the region co-spatial with the radio emission. Broad and blueshifted (~ 500 km s^{-1}) emission is observed in the region of the radio lobe, at the same location as the blueshifted H I absorption. The velocity of the ionised outflow is similar to the one found in H I (see Figure 8.1 right). The first-order correspondence between the radio and optical properties suggests that the outflow is driven by the interaction between the radio jet and the ISM.

Other cases of fast H I outflow have been found and summarised in Morganti *et al.* (2005); see also Figure 8.2. The main result of this study is that the neutral outflows occur, in at least some cases, at kpc distance from the nucleus and, as in the case of IC 5063, they are most likely driven by the interactions between the expanding radio jets and the gaseous medium enshrouding the central regions. We estimate that the

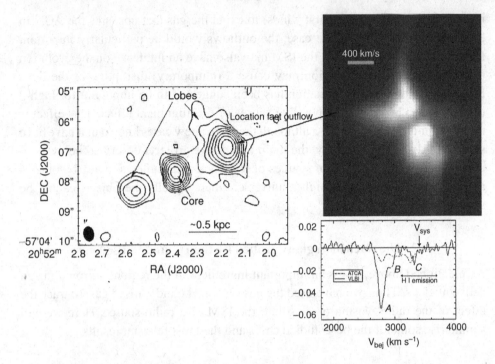

Figure 8.1 *Left:* Radio continuum image of IC 5063. *Right:* Comparison between the width of the H I absorption (bottom plot) and that of the ionised gas (top; from the [O III]5007Å). The first-order similarity between the amplitude of the blueshifted component is clearly seen.

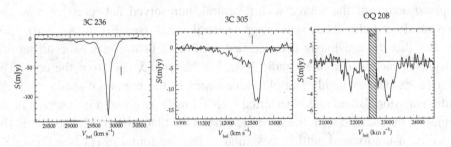

Figure 8.2 H I absorption profiles detected against three radio galaxies. The observations of the radio galaxies were done using the WSRT, see Morganti *et al.* (2005) for details. The short vertical line indicates the systemic velocity.

associated mass outflow rates are up to $\sim 50\ M_\odot\ \mathrm{yr}^{-1}$, comparable (although at the lower end of the distribution) to the outflow rates found for starburst-driven superwinds in ultra luminous IR galaxies, see Rupke *et al.* (2002). This suggests that massive, jet-driven outflows of neutral gas in radio-loud AGN can have as large an impact on the evolution of the host galaxies as the outflows associated with starbursts. This is important as starburst-driven winds are recognised to be responsible for inhibiting early star formation, enriching the ICM with metals and heating the ISM/IGM medium. In the case of IC 5063, a few more parameters related to the gaseous outflow could be derived. The mass outflow rates of cold (H I) and warm (ionised) gas have been found to be comparable, $\sim 30\ M_\odot\ yr^{-1}$. With a black-hole mass of $2.8 \times 10^8\ M_\odot$, the Eddington luminosity of IC 5063 is $3.8 \times 10^{46}\ erg\ s^{-1}$; this means that the kinetic power of the outflow represents only about few $\times 10^{-4}$ of the available accretion power. This result is similar to that found for PKS 1549-79 (see below). However, unlike the case of PKS 1549-79, IC 5063 accretes at a low rate ($\dot m \sim 0.02$). Thus, in IC 5063 the kinetic power of the outflow appears to be a relatively high fraction of the nuclear bolometric luminosity ($\sim 8 \times 10^{-2}$). In IC 5063, the observed outflows may have sufficient kinetic power to have a significant impact on the evolution of the ISM in the host galaxy.

8.2.2 Young radio sources: the case of PKS 1549-79

As mentioned above, gas-rich mergers have often been identified as one possible trigger for AGN activity. It is not clear how common this is, but certainly in some cases the black hole may grow rapidly through merger-induced accretion following the coalescence of the nuclei of two merging galaxies. If this is the case, the major growth phase is largely likely to happen hidden (at optical wavelengths) by the natal gas and dust. We have recently identified one relatively nearby system in such a phase of evolution – PKS1549-79 ($z = 0.152$). This object shows all the characteristics expected for a proto-quasar (see Holt *et al.* 2006 for details). These include a high accretion rate onto the supermassive black hole, a large reddening at optical wavelengths, evidence for rapid AGN-driven outflows in the warm emission line gas and morphological evidence that the activity has been triggered in a major galaxy merger. The signatures of this merger can be seen in tidal tails in the optical (see Figure 8.3 left) and by the presence of a substantial young stellar population (50–250 Myr).

PKS 1549-79 is also a small (~ 200 pc) and young but powerful radio galaxy. The radio source has a core-jet structure that is considered to be quite closely aligned with the line-of-sight. Although no broad permitted (optical) lines were detected, broad Paα in NIR and a reddened continuum spectrum were observed, indicating the presence of a hidden quasar nucleus in this source . The presence of a fast outflow is revealed by the large blueshifts ($\Delta V \sim 680$ km s^{-1}) and large line widths (FWHM ~ 1300 km s^{-1}) of the high ionisation optical emission lines (e.g. [OIII], [NeV]). H I absorption is surprisingly present, indicating that it must originate not from a circumnuclear disc but from gas in which the radio source is

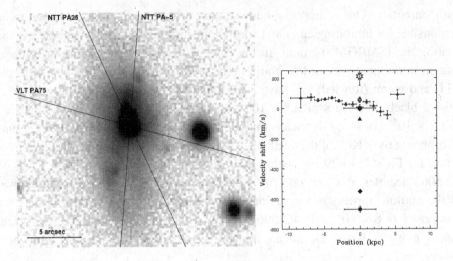

Figure 8.3 *Left:* Deep VLT r-image of PKS1549-79. The slit positions used for the long-slit spectroscopic observations are shown for reference. *Right:* Radial velocity profiles of PKS 1549-79 obtained from the slit position PA-5. Small open and filled triangles represent the narrow and intermediate components of Hα respectively. Overplotted is the radial velocity of the deep H I 21 cm absorption (large filled triangle at -30 km s^{-1}). More details can be found in Holt *et al.* (2006).

deeply embedded. This is suggested by the fact that the VLBI observations show that the H I is uniformly distributed across the radio source with no velocity or column density gradients. However, despite the evidence for rapid outflows in the warm gas, the estimated kinematic power in the warm outflow is several orders of magnitude less than required by the feedback models ($P_{kin}/L_{edd} < 10^{-4}$: Holt *et al.* 2006). One possible explanation for this apparent discrepancy is that much of the mass of the AGN-induced outflow is tied up in cooler or hotter phases of the ISM. Unlike in objects such as IC 5063, the search done so far to detect a broad absorption component of the H I has not been successful, and has been limited by the poor data quality. X-ray observations are also planned. It is interesting to compare these results with those obtained in the study of ionised gas in other young radio sources. The modelling of the complex kinematics of the gas in these sources shows that broad components are common and that they tend to be blueshifted compared to the systemic velocity of the host galaxy (derived from the extended and quiescent gas). Thus, these broad components have been interpreted as gas outflows, possibly driven by jet–ISM interaction (see Holt *et al.* 2008 for the full discussion). Comparisons with samples in the literature also show that compact, young radio sources harbour more extreme nuclear kinematics than their extended counterparts, a trend seen within the sample studied by Holt *et al.* (2008) with larger velocities in the smaller sources. The observed velocities are also likely to be influenced by source orientation with respect to the observer's line of sight.

8.3 Moving to larger scales: jet-induced star formation

Current theoretical models suggest that radio-source shocks propagating through the clumpy ISM/IGM trigger the collapse and/or fragmentation of overdense regions, which may then subsequently form stars (e.g. Fragile *et al.* 2004; Mellema *et al.* 2002, and references therein).

Jet-induced star formation is considered to be particularly important for high-*z* radio galaxies. Despite the many examples found in the nearby universe of gas shocked by the interaction with the radio jet, there are not many cases of jet-induced star formation known at low-*z*. Nevertheless, the few cases known are ideal for studying the details of such interaction. Here we discuss a few of these nearby examples.

8.3.1 Jet-induced star formation in Centaurus A

The north-east region of Centaurus A, located about 15 kpc from its nucleus, is a particularly complex site where different structures have been found and studied. In particular, an H I ring (Schiminovich *et al.* 1994) is situated (in projection) near the radio jet of Centaurus A, as well as near very turbulent filaments of highly ionised gas and near regions with young stars. This is illustrated in Figure 8.4 left. The spatial coincidence of these structures, together with the fact that the ionised and neutral gas cover the same velocity range, has led to the suggestion that the radio jet is interacting with the H I cloud, producing the turbulent gas filaments and *triggering the star formation in this region.*

The higher velocity and spatial resolution of ATCA H I data (Oosterloo & Morganti 2005) indeed reveal that, in addition to the smooth velocity gradient corresponding to the overall rotation of the H I gas around Centaurus A, H I with anomalous velocities of about 100 km s^{-1} is present at the southern tip of this cloud. This is interpreted as evidence for an ongoing interaction between the radio jet and the H I cloud. Gas stripped from the H I cloud gives rise to the large filament of ionised gas and the star formation regions that are found downstream from the location of the interaction. The implied flow velocities are very similar to the observed anomalous H I velocities. Given the amount of H I with anomalous kinematics and the current star formation rate, the efficiency of jet-induced star formation is, at most, of the order of one per cent. If this overall description is correct, the jet-induced star formation is fairly inefficient. Mould *et al.* (2000) report a star formation rate for the region of the order of a few times $10^{-3} \ M_\odot \text{ yr}^{-1}$.

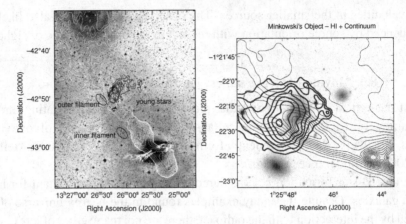

Figure 8.4 *Left:* Overlay showing the positions of the various components of Centaurus A. The optical image shows the well-known dust lane of Centaurus A and the faint diffuse optical emission that extends to very large radius. The white contours denote the radio continuum emission The black contours denote the H I cloud and the locations of the inner and outer filaments of highly ionised gas are indicated, as well as the location of young stars (from Oosterloo & Morganti 2005). *Right:* Map of the H I cloud associated with the MO, with radio continuum contours from the VLA overlaid (from Croft *et al.* 2006).

Assuming that the star formation rate has been constant, this implies a total mass for the stars formed over 15 Myr (the age of the young stars) to be of the order of a few times 10^4 $M_\odot$. The amount of H I showing the anomalous velocities is about 1×10^6 $M_\odot$. Thus, unless the current rate by which the H I is stripped from the cloud is much higher than in the past, this appears to imply that the efficiency of converting the gas stripped off the cloud into stars is at most a few per cent.

8.3.2 Starburst triggered by a radio jet in the Minkowsky object

Multi-waveband data have shown new evidence that the star formation in the Minkowsky object (MO), a star-forming peculiar galaxy near NGC 541, was induced by a radio jet (Croft *et al.* 2006). Key findings are the discovery of a 4.9×10^8 $M_\odot$ double H I cloud straddling the radio jet downstream from the MO, where the jet changes direction and de-collimates (see Figure 8.4 right).

This is similar to the jet-induced star formation associated with the Centaurus A jet, and the radio-aligned star-forming regions in powerful radio galaxies at high redshift. The age of the MO has been estimated to be around 7.5 Myr. While it is not possible to completely rule out the presence of an old population in the MO, the data are consistent with the MO having formed *de novo* when the jet interacted with the ambient ISM/IGM. Unlike Centaurus A, we propose that the

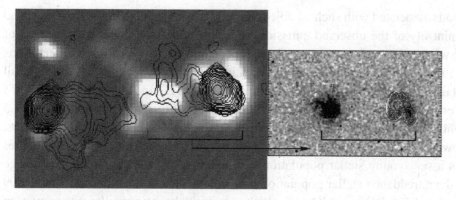

Figure 8.5 *Left:* Low-resolution ATCA radio image superimposed on narrow-band image (Clark *et al.* 1997). *Right:* High-resolution WFPC2 F547M continuum image of PKS 2250-41 overlaid with 15 GHz VLA radio contours (from Inskip *et al.* 2008). The optical (as well as the infrared) image displays flux at the position of peak emission line intensity in the western arc, which is also coincident with the secondary hotspot within the western radio lobe.

H I associated with the MO formed *in situ* through cooling of clumpy, warm gas in the stellar bridge or cluster IGM, as it was compressed by radiative shocks at the jet collision site, in agreement with numerical simulations (Fragile *et al.* 2004). The star formation in the MO then followed from the cooling and collapse of such H I clouds, and the H I kinematics, which show 40 km s^{-1} shear velocities, are also consistent with such models.

8.3.3 A new candidate: PKS 2250-41

PKS 2250-41 is an archetypal example of a galaxy displaying jet–cloud interactions, with clear evidence for shocks associated with the expanding radio source. This is illustrated by the observed distribution of ionised gas (see Figure 8.5 left) and by the variation in ionisation state and gas kinematics in the vicinity of the western radio lobe hotspots. Past studies of PKS2250-41 have suggested that the prominent emission line arc to the west of the host galaxy originates from a direct collision between the radio source jet and a companion galaxy (Clark *et al.* 1997; Villar-Martí *et al.* 1999). The primary evidence is the continuum emission which, in addition to the well-studied line emission, is also observed within the western radio lobe. The continuum emission is approximately co-spatial with the secondary radio source hotspot (Figure 8.5 right), has low polarisation and only a limited contribution of nebular continuum emission. Clark *et al.* (1997) suggested that the residual continuum emission could originate from a late-type spiral or irregular galaxy, with which the radio jet has collided; radio source shocks driven through the gas clouds associated with such an object can also account for the impressive scale and luminosity of the observed emission line arc, and possibly also the shortness of the western lobe relative to the eastern lobe. New optical and infrared observations of PKS 2250-41 add further weight to this scenario (see Inskip *et al.* 2008 for all details). Continuum emission is detected in both the K_S and on the WFPC2 F547M filters within the arc, coincident with the secondary radio hotspot. Figure 8.5 right displays the high-resolution optical continuum overlaid with radio contours. However, and interestingly, the inferred spectral shape of the continuum implies that a *very* young stellar population is dominating the optical emission. The age of the unreddened stellar population has been estimated to be either 0.006–0.009 Gyr or 0.05–0.1 Gyr, or alternatively the age could be even smaller if the emission is produced by highly reddened young stellar populations (YSP). This suggests that the proximity of the radio source may very well have triggered recent star formation within this object.

8.4 Conclusions

Clear evidence of the impact of the interaction between the radio jets and the ISM has been found from the study of the kinematics of the gas in radio-loud sources. Fast outflows of neutral atomic hydrogen and ionised gas are produced by such interaction. The presence of neutral gas associated with such outflows indicates that the gas can cool very efficiently following a strong jet–cloud interaction. Outflows of similar velocities are observed in H I and in ionised gas, but the mass outflow rate is relatively high in H I and much lower in ionised gas. The derived mass outflow rate in H I ranges between a few and ~ 50 $M_\odot$/yr, comparable (although at the lower end) to that found in ultraluminous IR galaxies. Thus, jet-driven outflows can have a similar impact on the evolution of a galaxy as starburst-driven superwinds. However, the importance for the feedback is not completely clear. In IC 5063 the outflow energy is a reasonable fraction of the nuclear bolometric luminosity (but not of the Eddington luminosity). On the other hand, PKS 1549-79 is in a stage where the nucleus is still hidden (in the optical) by the gas/dust coming from the merger that triggered the radio source. However, the outflow of ionised gas is not as large as expected in the quasar feedback model (while H I outflow has not been found yet in this radio source). Thus, this study so far indicates that outflows of ionised gas are typically not massive enough to clear the nuclear gas in young radio sources, while the situation is more promising for outflows of cold gas, although at the moment the statistics on occurrence of such outflows are limited. This result appears to be confirmed for other objects (see Tadhunter 2008 for an overview).

Jet-induced star formation has been found in a very limited number of nearby radio galaxies. We have presented here a possible new case. The jet-induced star formation appears to be relatively inefficient both in the case of Centaurus A and in the MO. Comparing the global star formation efficiency $M_{stars}/M_{H\,I}$ in MO we found a value of $\sim 4\%$, which is similar to that in Centaurus A (Oosterloo & Morganti 2005). However, different origins have been suggested for the two systems. *In situ* formation through cooling of clumpy, warm gas has been suggested in the case of the Minkowsky object, while in the case of Centaurus A the H I was likely already present as large-scale structure. Unfortunately, no H I information is available for PKS 2250-41 because the system is at too high a redshift.

In summary, H I and ionised gas observations of radio-loud sources provide extra constraints on the effects of this kind of AGN on the ISM. Although we are still building up the statistics for a large number of objects, these studies are particularly important in order to get a more complete and realistic picture of the effects of feedback in galaxy evolution.

Acknowledgements

The author would like to thank the organisers of the very interesting workshop for their kind invitation. The author would also like to acknowledge and thank the main collaborators involved in the projects presented here: Clive Tadhunter, Tom Oosterloo, Joanna Holt, Katherine Inskip, Steve Croft and Wil van Breugel.

References

Best, P. N., Kauffmann, G., Heckman, T. M., *et al.* (2005) The host galaxies of radio-loud active galactic nuclei: mass dependences, gas cooling and active galactic nuclei feedback. *MNRAS*, **362**, 25

Clark, N. E., *et al.* (1997) Radio, optical and X-ray observations of PKS 2250-41: a jet/galaxy collision? *MNRAS*, **286**, 558

Combes, F., Young, L. M., Bureau, M. (2007) Molecular gas and star formation in the SAURON early-type galaxies. *MNRAS*, **377**, 1795

Croft, S., *et al.* (2006) Minkowsky's Object: a starburst triggered by a radio jet, revised. *ApJ*, **647**, 1040.

Emonts, B. H. C., Morganti, R., Tadhunter, C. N., *et al.* (2006) Timescales of merger, starburst and AGN activity in radio galaxy B2 0648+27. *A&A*, **454**, 125

Fragile, P. C., Murray, S. D., Anninos, P., van Breugel, W. (2004) Radiative shock-induced collapse of intergalactic clouds. *ApJ*, **604**, 74

Holt, J., *et al.* (2006) The co-evolution of the obscured quasar PKS 1549-79 and its host galaxy: evidence for a high accretion rate and warm outflow. *MNRAS*, **370**, 1633

Holt, J., Tadhunter, C. N., Morganti, R. (2008) Fast outflows in compact radio sources: evidence for AGN-induced feedback in the early stages of radio source evolution. *MNRAS*, **387**, 639

Inskip, K. J., Villar-Martí, M., Tadhunter, C. N., *et al.* (2008) PKS 2250-41: a case study
for triggering. *MNRAS*, **386**, 1797

Mellema, G., Kurk, J., Rottgering, H. (2002) Evolution of clouds in radio galaxy cocoons.
A&A, **395L**, 13

Morganti, R., Tadhunter, C. N., Oosterloo, T. (2005) Fast neutral outflows in powerful
radio galaxies: a major source of feedback in massive galaxies. *A&A*, **444**, L9

Morganti, R., De Zeeuw, T., Oosterloo, T., *et al.* (2006) Neutral hydrogen in nearby
elliptical and lenticular galaxies: the continuing formation of early-type galaxies.
MNRAS, **371**, 157

Morganti, R., Holt, J., Saripalli, L., Oosterloo, T., Tadhunter, C. (2007) IC 5063: AGN
driven outflow of warm and cold gas. *A&A*, **476**, 735

Mould, J. R., *et al.* (2000) Jet-induced star formation in Centaurus A. *ApJ*, **536**, 266

Oosterloo, T. A., Morganti, R. (2005) Anomalous H I kinematics in Centaurus A:
Evidence for jet-induced star formation. *A&A*, **429**, 469

Rupke, D. S., Veilleux, S., Sanders, D. B. (2002) Keck absorption-line spectroscopy of
galactic winds in ultraluminous infrared galaxies. *ApJ*, **570**, 588

Sarzi, M., *et al.* (2006) The SAURON project – V. Integral-field emission-line kinematics
of 48 elliptical and lenticular galaxies. *MNRAS*, **366**, 1151

Schiminovich, D., van Gorkom, J. H., van der Hulst, J. M., Kasow, S. (1994) Discovery of
neutral hydrogen associated with the diffuse shells of NGC 5128. *ApJL*, **423**, 101

Silk, J., Rees, M. J. (1998) Quasars and galaxy formation. *MNRAS*, **331**, L1

Stanghellini, C., O'Dea, C. P., Dallacasa, D., *et al.* (2005) Extended emission around GPS
radio sources. *A&A*, **443**, 891

Tadhunter, C. N. (2008) The importance of sub-relativistic outflows in AGN host galaxies.
MemSAIt, **79**, 1205

Villar-Martí, M., Tadhunter, C., Morganti, R., Axon, D., Koekemoer, A. (1999) PKS
2250-41 and the role of jet–cloud interactions in powerful radio galaxies. *MNRAS*,
307, 24

9
Young radio sources: evolution and broad-band emission

L. Ostorero, R. Moderski, Ł. Stawarz, M. Begelman, A. Diaferio,
I. Kowalska, J. Kataoka & S. J. Wagner

9.1 Introduction

Radio galaxies represent ideal laboratories to investigate the triggering, maintenance, and fading of AGN activity, as well as the link between these processes and the physical conditions of the environment, from sub- to super-galactic scales. In this context, key sources for the study of the very first phases of the evolution of radio galaxies are the gigahertz-peaked-spectrum (GPS) sources associated with galaxies. From sub-kpc scales, their jet-lobe structures propagate through the host-galaxy interstellar medium (ISM), evolving into sub-galactic (compact-steepspectrum, CSS) sources, which then expand to super-galactic scales (see O'Dea 1998 for a review). However, this scenario still has several open issues, such as the absorption mechanism responsible for the characteristic turnover in the radio spectrum, the details of the dynamical evolution and interaction with the ISM, the parameters of the central engine, and the origin of the high-energy emission. We recently proposed a model that addresses some of these issues through the analysis of the broad-band emission of GPS galaxies (Stawarz *et al.* 2008). Here we show that our model satisfactorily reproduces a number of observed properties of X-ray emitting GPS galaxies.

9.2 The model: dynamical and spectral evolution

In the following, we recall the main features of our dynamical-radiative model; a more comprehensive discussion can be found in our original paper (Stawarz *et al.* 2008), and references therein. Our description of the dynamical evolution of GPS sources is based upon the model proposed by Begelman and Cioffi (1989) to explain the expansion of classical double sources in an ambient medium with

AGN Feedback in Galaxy Formation, eds. V. Antonuccio-Delogu and J. Silk. Published by Cambridge University Press. © Cambridge University Press 2011.

density profile $\rho(r)$. The relevant equations can be derived by assuming that: (1) the jet momentum flux (proportional to the jet kinetic power L_j) is balanced by the ram pressure of the ambient medium spread over an area A_h; (2) the sideways expansion velocity of the lobes equals the speed of the shock driven by the overpressured cocoon, with internal pressure p, in the surrounding medium; and (3) the energy $L_j t$ transported by the pair of jets during the source lifetime is converted into the cocoon's internal pressure. For young GPS sources with age t, linear size $LS(t) \lesssim 1$ kpc, and transverse size $l_c(t)$, expanding in the central core of the gaseous halo of the host galaxy, we could constrain the model with a number of reasonable assumptions: (1) a constant ambient density $\rho = m_p n_0$ (with m_p the proton mass, and $n_0 \simeq 0.1 \mathrm{cm}^{-3}$), representative of the inner core; (2) a constant hot-spot advance velocity $v_h \simeq 0.1c$, as suggested by many observations of compact symmetric objects; (3) a scaling law $l_c(t) \sim t^{1/2}$, reproducing the initial, ballistic phase of the jet propagation. Therefore, all the lobes' physical quantities become functions of two parameters only: the jet kinetic power L_j and the source linear size LS. We then studied how the broad-band radiative output of GPS sources evolves as the source expands, for a given jet power L_j. The electron population $Q(\gamma)$ (with γ the electron Lorentz factor), injected from the terminal jet shocks to the expanding lobes, evolves under the joint action of adiabatic and radiative energy losses, yielding a lobe electron population $N_e(\gamma)$, which has a broken power-law form with critical energy γ_{cr} when $Q(\gamma)$ is a power law, and a more complex form when $Q(\gamma)$ is a broken power law with intrinsic break $\gamma_{int} \simeq 2 \times 10^3$ (Stawarz *et al.* 2008). The lobe electrons are a source of synchrotron radiation, with luminosity L_{syn}.

Free–free absorption (FFA) of this radiation by neutral-hydrogen clouds of the narrow-line region (NLR), engulfed by the expanding lobes and photoionised on their external layer by the radiation from the active nucleus (as proposed by Begelman 1999), is the process that we favour for the formation of the inverted spectra. Whereas the synchrotron-self-absorption (SSA) process enabled us to reproduce neither the observed turnover frequencies v_p, nor the spectra below the turnover, nor the $v_p - LS$ anti-correlation (O'Dea & Baun 1997), FFA effects proved to best fit the inverted spectra, and are a promising candidate to account for the above anti-correlation.

The lobe particles are also a source of inverse-Compton (IC) radiation via up-scattering of both the synchrotron radiation (synchrotron-self-Compton mechanism; SSC) and the local, thermal photon fields generated by the accretion disc, the torus, and the stellar population of the host galaxy (external-Compton mechanism). The IC scattering of all the above radiation fields yields significant and complex high-energy emission, from the X-ray to the γ-ray energy domain. Whereas in GPS *quasars* the direct X-ray emission of the accretion disc's hot corona and of the beamed relativistic jets may overcome the X-ray output of most of these sources, in GPS *galaxies* those contributions are expected to be obscured by the torus and Doppler-hidden, respectively, and the lobes are expected to be the dominant X-ray source.

9.3 Comparison with broad-band spectra of GPS galaxies

Our model is a powerful tool to study the evolution of a typical GPS spectrum as a function of the time-dependent source linear size $LS(t)$, given the jet kinetic power L_j and the energy spectrum of the hot-spot electrons injected into the lobes; assuming typical luminosities for the putative torus and accretion-disc components, it is also possible to investigate the temporal evolution of the high-energy, comptonised component of the spectral energy distribution (SED).

By applying the model to sources with *measured* linear sizes, we could constrain their jet kinetic powers and the spectra of their hot-spot particles, test the viability of the FFA effect as the main effect responsible for the optically thick part of the radio spectra, and evaluate the contribution of the comptonised radiation to the high-energy emission of the source.

The source sample we chose to apply our model to is the sample of 11 GPS galaxies known as X-ray emitters up to 2008.[1] The spectral modelling of the complete sample is presented in Stawarz *et al.* (2008). In Figure 9.1, we show, as an example, the modelling of the intrinsic radio spectrum and broad-band SED of B0108+388, a source with $LS = 41$ pc.[2] The broad-band data were derived from the literature, and properly de-absorbed. The SED of IERS B0108+388 was modelled by associating a kinetic power $L_j = 2.05 \times 10^{45}$ erg s^{-1} to the source jets. Synchrotron radiation produced by a lobe electron population $N_e(\gamma)$ could reproduce the radio data at frequencies above the turnover. $N_e(\gamma)$ was derived from the evolution of an injected hot-spot population $Q(\gamma) \sim \gamma^{-s}$, with $s = s_1 = 1.8$ for $\gamma < \gamma_{int}$, and $s = s_2 = 3.2$ for $\gamma > \gamma_{int}$. FFA effects enabled us to best fit the spectral behaviour at frequencies below the ~ 6 GHz turnover. The thermal emissions from the torus (IR), the disc (UV), and the host galaxy (optical-NIR) were modelled as black-body spectra with the appropriate frequency peaks ($\nu_{IR} = 0.5 \times 10^{13}$ Hz, $\nu_{UV} = 2.45 \times 10^{15}$ Hz, and $\nu_{opt} = 2.0 \times 10^{14}$ Hz) and bolometric luminosities ($L_{IR} = 5.0 \times 10^{44}$ erg s^{-1}, $L_{UV} = 5.0 \times 10^{45}$ erg s^{-1}, and $L_{opt} = 6.0 \times 10^{44}$ erg s^{-1}). The comptonisation of the synchrotron and thermal radiation

---

[1] IERS B0026+346, IERS B0108+388*, IERS B0500+019*, IERS B0710+439, PKS B0941-080, IERS B1031+567*, IERS B1345+125*, IVS B1358+624*, IERS B1404+286, IERS B2128+048, IERS B2352+495*. Sources marked with an asterisk are those of the subsample discussed in Section 9.4, and included in Figure 9.2.

[2] Throughout this paper, we use the cosmological parameters: $\Omega_\Lambda = 0.7$, $\Omega_M = 0.3$, with $H_0 = 72$ km s^{-1}Mpc^{-1}.

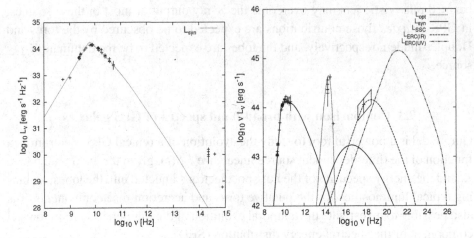

Figure 9.1 Modelling of the intrinsic radio spectrum (*left panel*) and broad-band SED (*right panel*) of GPS galaxy IERS B0108+388. For the sources of radio to X-ray data, see Ostorero *et al.* (2009). The curves show the modelled spectral components: synchrotron emission, and corresponding SSC emission (solid lines); thermal star light (dotted line); comptonised thermal emission from the torus and the disc, respectively (dashed and dash-dot-dotted lines). The comptonised starlight's luminosity does not appear in the plot because it is lower than 10^{42} erg s^{-1}.

fields yielded the high-energy spectral components. For this source, the X-ray emission is dominated by the comptonisation of the IR radiation. Our model well reproduces the observed X-ray spectrum, and predicts significant γ-ray emission.

9.4 Further observational support

The prediction of X-ray emitting lobes in GPS galaxies, which is one of the distinctive features of our model, may be supported by further observational evidence, as discussed in Ostorero *et al.* (2009).

The traditional interpretation of the GPS X-ray emission as thermal radiation from the accretion disc, absorbed by an AGN gas component with equivalent hydrogen column density N_H, was mainly based on the apparent discrepancy between the equivalent total-hydrogen column density N_H derived from the X-ray spectral analysis and the neutral-hydrogen column density N_{HI} derived from the 21-cm radio measurements. Because N_H always appeared to exceed N_{HI} by 1–2 orders of magnitude, it became natural to interpret the X-rays as produced in a source region that is more obscured than the region where the bulk of the radio emission comes from, and thus located *in between* the radio lobes; otherwise, an unreasonably high fraction of ionised hydrogen (H II) would be necessary to account for the above difference (e.g. Guainazzi *et al.* 2006; Vink *et al.* 2006). Such a scenario would also be consistent with the observed anti-correlation between N_{HI} and linear size found by Pihlström *et al.* (2003), the fraction of ionised gas likely being low in a young radio source with a still expanding Strömgren sphere (Vink *et al.* 2006).

However, the discrepancies between the N_H and N_{HI} values mentioned above should be regarded with caution. The N_{HI} estimate is derived, from the measurements of the H I absorption lines, as a function of the ratio between the spin temperature T_s of the gas and its covering factor c_f, representing the fraction of the source covered by the H I screen. The common assumption $T_s/c_f = 100$ K refers to the case of complete coverage ($c_f = 1$) of the emitting source by a standard cold ($T_k \simeq 100$ K) ISM cloud in thermal equilibrium, and thus with spin temperature equal to the kinetic temperature ($T_s = T_k$). However, this assumption returns a value of N_{HI} that represents a lower limit to the actual neutral hydrogen column density (e.g. Pihlström *et al.* 2003; Vermeulen *et al.* 2003). In fact, in the AGN environment, illumination by X-ray radiation might easily raise T_k to $10^3 - 10^4$ K (e.g. Maloney *et al.* 1996), making T_s rise accordingly (e.g. Listz 2001); a source covering factor smaller than unity would also increase the T_s/c_f ratio. Both the above effects might lead to N_{HI} values fully consistent with the N_H estimates. Finally, temperatures as high as several 10^3 K would likely imply the presence of a non-negligible fraction of H II (Maloney *et al.* 1996; Vink *et al.* 2006), also contributing to relax possible residual column-density discrepancies. The consistency of N_H and N_{HI} would make the scenario of non-thermal X-ray-emitting lobes a viable alternative to the accretion-disc dominated model.

A way to unveil the actual X-ray production site is to compare the properties of the X-ray and radio absorbers, i.e. the N_H and N_{HI} column densities. Such a comparison can be performed either for individual sources, where an ad hoc increase of the T_s parameter can remove possible N_H and N_{HI} discrepancies, or for a source sample, where the existence of a positive, significant N_H–N_{HI} correlation would suggest that the X-ray and radio absorbers coincide, thus supporting the co-spatiality of the X-ray and radio source.

We investigated the existence of a connection between N_H and N_{HI} in our GPS sample. For a positive correlation, we searched the source subsample for which both N_H and N_{HI} estimates (either detections or upper limits) are available (see Figure 9.2, and footnote 1). We obtained the following results: (1) the subsample of five sources for which both N_H and N_{HI} *detections* are available displays a strong (Pearson's $r = 0.997$) and highly significant ($S = 2.3 \times 10^{-4}$) N_H–N_{HI} positive correlation; (2) in the above-mentioned subsample, the strength and significance of the correlation substantially decrease when using non-parametric methods (Kendall's and Spearman's correlation coefficients); (3) the six-source subsample

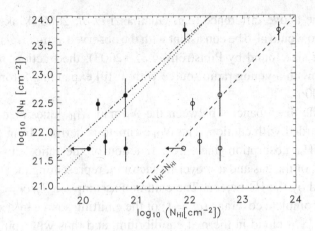

Figure 9.2 N_H vs. N_{HI} for six GPS of our sample. Solid symbols: N_{HI} was computed with $T_s = 100$ K; arrows are upper limits. Open symbols: as an example, the same sources with $T_s = 6 \times 10^3$ K are shown. Dash-dotted line: linear fit to the five-source subsample of N_H/N_{HI} *detections* (with $T_s = 100$ K); dotted line: linear fit to the six-source subsample including both *detections* and *upper limits*. For the data sources, see Ostorero *et al.* (2009).

including the N_{HI} *upper limit* also shows, according to survival analysis techniques (ASURV, Rev. 1.2; e.g. Lavalley *et al.* 1992), a positive correlation, however with lower strength ($\rho \sim 0.6$) and a significance varying in the range $S = 0.17-0.33$, depending on both the N_H value (when more than one is available) and the statistical method chosen for the analysis. Although the data are suggestive of a positive correlation, further measurements would definitely help to improve the statistics.

9.5 Conclusions and future prospects

Our dynamical-radiative model can reproduce the observed broad-band SED of X-ray emitting GPS galaxies. The shape of the synchrotron radio spectra at frequencies lower than the turnover is best accounted for by assuming FFA effects as the dominant absorption mechanism. The X-ray spectra can be ascribed to IC scattering of the thermal radiation fields (accretion disc, torus and host galaxy) off the lobes' electron population. Independent observational support for the scenario of X-ray emitting lobes comes from the radio and X-ray hydrogen column densities of a sample of X-ray GPS: the data are suggestive of a positive correlation, which, if confirmed, would point towards the co-spatiality of the radio and X-ray emission sites. We have been performing additional measurements, necessary to improve the statistics.

Acknowledgements

L.O. and A.D. acknowledge partial support from the INFN grant PD51. Ł.S., R.M., J.K. and S.W. acknowledge support of MEiN grant 1-P03D-003-29, MNiSW grant N N203 301635, JSPS KAKENHI (19204017/14GS0211) and BMBF/DLR grant 50OR0303, respectively. We wish to thank the organisers for a fruitful and successful meeting.

References

Begelman, M. C. and Cioffi, D. F. (1989). Overpressured cocoons in extragalactic radio sources. *Astroph. J.*, **345**, L21–L24

Begelman, M. C. (1999). Young radio galaxies and their environment, in *The Most Distant Radio Galaxies*, eds. H. J. A. Röttgering, P. N. Best and M. D. Lehnert (Amsterdam Royal Netherlands Academy of Arts and Sciences)

Guainazzi, M., Siemiginowska, A., Stanghellini, C. *et al.* (2006). A hard X-ray view of giga-hertz peaked spectrum radio galaxies. *Astron. Astrophys.*, **446**, 87–96

Lavalley, M. P., Isobe, T. and Feigelson, E. D. (1992). ASURV, Pennsylvania State University. Report for the period Jan 1990 – Feb 1992. *Bull. Am. Astron. Soc.*, **24**, 839–840

Listz, H. (2001). The spin temperature of warm interstellar HI. *Astron. Astrophys.*, **371**, 698–707

Maloney, P. R., Hollenbach, D. and Tielens, A. G. G. M. (1996). X-ray irradiated molecular gas. I. Physical processes and general results. *Astroph. J.*, **466**, 561–584

O'Dea, C. P. (1998). The Compact Steep-Spectrum and Gigahertz Peaked-Spectrum radio sources. *Publ. Astron. Soc. Pac.*, **110**, 493–532

O'Dea, C. P. and Baum, S. A. (1997). Constraints on radio source evolution from the Compact Steep Spectrum and GHz Peaked Spectrum radio sources. *Astron. J.*, **113**, 148–161

Ostorero, L., Moderski, R., Stawarz, Ł. *et al.* (2009). Modelling the broad-band spectra of X-ray emitting GPS galaxies. *Astron. Nachr.*, **330**, 275–278

Pihlström, Y. M., Conway, J. E. and Vermeulen, R. C. (2003). The presence and distribution of HI absorbing gas in sub-galactic sized radio sources. *Astron. Astrophys.*, **404**, 871–881

Stawarz, Ł., Ostorero, L., Begelman, M. C. *et al.* (2008). Evolution of and high-energy emission from GHz-peaked spectrum sources. *Astroph. J.*, **680**, 911–925

Vermeulen, R. C., Pihlström, Y. M., Tschager, W. *et al.* (2003). Observations of HI absorbing gas in compact radio sources at cosmological redshifts. *Astron. Astrophys.*, **404**, 861–870

Vink, J., Snellen, I., Mack, K.-H. and Schilizzi, R. (2006). The X-ray properties of young radio-loud AGN. *Mon. Not. R. Astron. Soc.*, **367**, 928–936

10

The duty cycle of radio galaxies and AGN feedback

S. Shabala

10.1 Introduction

In recent years, remarkably tight correlations have been observed between the properties of supermassive black holes (SMBHs) residing in galaxy cores and those of the host galaxies themselves (Magorrian *et al.* 1998; Gebhhardt *et al.* 2000; Häring and Rix 2004). A growing body of evidence seems to support the idea that feedback from active galactic nuclei (AGN) provides a natural link between these. While every galaxy can potentially host a SMBH, only a relatively small fraction of these are observed in an active state. AGN activity manifests itself through powerful outflows observable right across the electromagnetic spectrum.

The central black holes are powered by accretion of surrounding cold gas. The resultant outflows, in turn, affect the cold gas supply by heating and/or transporting this gas away from dense inner regions with short cooling times. It is for this reason that feedback from radio sources is particularly interesting. Despite only contributing around one per cent of the AGN bolometric luminosity, radio-loud AGNs can profoundly affect their surroundings through such mechanical feedback. One piece of observational evidence supporting this view comes from studies of X-ray clusters. In the absence of feedback, large amounts of cold gas are expected in dense cluster cores (due to short cooling times), however no such gas has been found. This well-known 'cooling flow problem' points to the need for a central, powerful heating source. Most cD galaxies contain radio sources at their centres (Burns 1990), and in many cases strong interaction between radio sources and their environment has been observed, with overpressured radio lobes displacing the hot cluster gas as they expand (Fabian *et al.* 2003; Forman *et al.* 2004). Shabala and Alexander (2009a) showed that intermittent AGN jets can suppress cooling flows while reproducing the observed ICM density structure in such clusters. An

AGN Feedback in Galaxy Formation, eds. V. Antonuccio-Delogu and J. Silk. Published by Cambridge University Press. © Cambridge University Press 2011.

even bigger problem alleviated by AGN feedback relates to number counts of the brightest galaxies. Standard Press–Schechter formalism overpredicts counts at both the bright and faint ends of the optical luminosity function. While feedback from supernovae resolves the discrepancy at the faint end, it is insufficient for the bright end. Observations indicate that galaxy formation is anti-hierarchical (the so-called 'cosmic downsizing' (Cowie *et al.* 1996)), with the most massive galaxies preferentially forming their stars early and having the bulk of their star formation quenched at $z \sim 1$. By contrast, star formation in-low mass galaxies proceeds until the present epoch.

AGN feedback is an inherently intermittent, self-regulating process. Fuel (in the form of cold gas) is required to power the outflows, and the outflows themselves limit cold gas availability. One could envisage an equilibrium-type configuration in which the heating and cooling rates were balanced; however, the effects of dynamical processes such as inflows, mergers and various instabilities make this unlikely. It therefore makes sense to talk about a duty cycle of AGN activity, and 'on' and 'off' timescales associated with this intermittency.

In this contribution I discuss how combining observations of local galaxies together with detailed radio source models can constrain AGN timescales, and how these depend on host galaxy properties. The physical interpretation of these results then allows us to successfully add a physically motivated prescription for radio source feedback to galaxy formation models.

10.2 Local sample

Qualitatively, the power of an AGN radio jet depends on the black hole accretion rate. This, in turn, is related to the rate of gas cooling and accretion disk properties. For the same type of disk, a greater fuelling rate corresponds to higher jet power. The radio jets are liable to disruption by Rayleigh–Taylor and Kelvin–Helmholtz instabilities. The Rayleigh–Taylor instability occurs when the underdense (with respect to the ambient medium) cocoon of radio plasma is decelerated at large radii so much that it can no longer support the heavier ambient gas against gravity. In the absence of extreme events, the shear (Kelvin–Helmholtz) instability is only likely to disrupt the jet as it propagates through the dense central regions of the galaxy. As discussed in detail below, dynamical models of radio source evolution suggest that both the expansion rate and maximum stable jet length are functions of the ambient gas density profile and jet power. In order to get a handle on these quantities, it is therefore crucial to consider radio properties of the AGNs together with optical properties of their host galaxies.

Shabala *et al.* (2008) constructed a radio/optical sample by cross-correlating the optical SDSS Data Release 2 (York *et al.* 2000) catalogue with the VLA FIRST

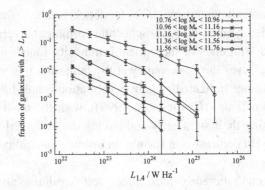

Figure 10.1 Cumulative fraction of AGNs brighter than a specified radio luminosity, as a function of stellar mass.

(Becker *et al.* 1995) and NVSS (Condon *et al.* 1998) 1.4 GHz radio surveys. A similar approach was taken by Best *et al.* (2005a). The main difference between the two catalogues is that the present one is complete at both the radio and optical wavelengths. The final flux- and volume-limited sample consists of 1191 radio sources at $0.03 \le z < 0.1$ with optical information.

10.2.1 Radio-loud AGN fractions

The radio/optical sample includes both star-forming galaxies (in which the radio emission arises from supernovae-driven shocks) and radio AGNs. These can be separated using a number of diagnostics. Perhaps the best known of these is the so-called BPT diagram (Baldwin *et al.* 1981), based on the emission line properties of the objects as given by their location in the [OIII] 5007/Hβ-[NII] 6583/Hα plane. Best *et al.* (2005a) introduced a radio flux-based classification by considering the predicted contribution to 1.4 GHz radio luminosity from supernovae remnants. The 4000 Å break, $D_n(4000)$, provides an indicator of stellar age, and together with the knowledge of stellar mass, $M_\star$, the location of a galaxy in the $L_{1.4}/M_\star - D_n(4000)$ plane determines whether the radio emission has a substantial AGN component. This was the criterion used to select radio-loud AGNs in the presented sample.

Figure 10.1 shows the fraction of galaxies brighter than a given radio luminosity as a function of galaxy mass. The value for the lowest luminosity bin corresponds to the radio-loud AGN fraction at the stellar mass of interest. Just as in Best *et al.* (2005b), a strong dependence of the radio-loud fraction on galaxy stellar mass is found, with the radio-loud fraction increasing by a factor of $\sim$30 for less than an order of magnitude rise in stellar mass. This suggests that massive galaxies spend a larger fraction of their lives in the radio active phase than do their lower-mass

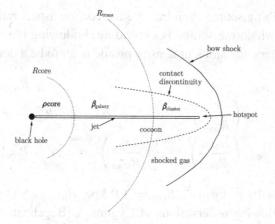

Figure 10.2 Radio source and environmental parameters.

counterparts. The question is, does this occur because they are 'on' for longer, 'off' for less time, or both?

10.2.2 P–D tracks

For a homogeneous class of galaxies that spend t_{on} in the active state and t_{off} in the quiescent state, the radio-loud fraction only constrains the ratio t_{on}/t_{off}. While the luminosity of a radio source evolves appreciably as it ages, unfortunately radio luminosity functions are not particularly sensitive to changes in source age (Shabala *et al.* 2008). However, combining luminosity and size information allows an estimate of both radio source age and jet power via the use of a dynamical model.

The model employed here was developed by Kaiser and Alexander (1997) and Kaiser *et al.* (2007). Accretion onto the black hole produces an initially ballistic jet. The jet is collimated by reconfinement shocks and is in pressure equilibrium with the surrounding material. This surrounding material is in fact a cocoon of radio plasma, inflated by backflow down the pressure gradient from the site of jet termination, the hotspot. The cocoon is overpressured with respect to the ambient gas, and therefore expands. The expansion is supersonic, driving a bow shock through the gas. The gas in the path of the radio source is thus shock heated and swept up in a snowplough-like manner (Figure 10.2).

The Kaiser and Alexander model predicts both source sizes (equivalent to twice the jet length) and synchrotron radio luminosities. The latter are derived under the assumption of equipartition between the magnetic field and thermal energy densities, with adiabatic, inverse-Compton (off the CMB photons) and synchrotron losses accounted for (Kaiser *et al.* 2007). The model predicts self-similar source evolution, with the cocoon aspect ratio R_T remaining constant. The important parameters describing source evolution are the cocoon aspect ratio, jet power and atmosphere into which the source is expanding. Following observations of local galaxies and clusters, the ambient density profile is set to be a double power-law,

$$
\rho(r) = \begin{cases} \rho_{\text{core}} & , r \leq R_{\text{core}}, \\ \rho_{\text{core}} \left(\dfrac{r}{R_{\text{core}}} \right)^{-\beta_{\text{galaxy}}} & , R_{\text{core}} < r \leq R_{\text{trans}}, \\ \rho_{\text{core}} \left(\dfrac{R_{\text{trans}}}{R_{\text{core}}} \right)^{-\beta_{\text{galaxy}}} \left(\dfrac{r}{R_{\text{trans}}} \right)^{-\beta_{\text{cluster}}} & , r > R_{\text{trans}}, \end{cases} \quad (10.1)
$$

where $\rho_{\text{core}} = 2 \times 10^{-22}$ kg m^{-3}, $R_{\text{core}} = 1.0$ kpc, $R_{\text{trans}} = 50$ kpc, $\beta_{\text{galaxy}} = 1.0$, $\beta_{\text{cluster}} = 1.9$. Following observations of Cygnus A (Begelman and Cioffi 1989), the cocoon axial ratio is set to be $R_T = 2.0$. The remaining parameters affecting source size and luminosity are jet power Q_{jet} and source age t_{on}. Given an atmosphere and cocoon aspect ratio, the tracks in the radio power–size (P–D) plane do not intersect, allowing jet powers and source ages to be reconstructed from observed source sizes and luminosities. Figure 10.3 shows the observed distributions in the P–D plane as a function of stellar mass.

10.2.3 Jet powers and timescales

It is clear from Figure 10.3 that the most massive galaxies host a higher fraction of larger, brighter sources. Figure 10.4 and Table 10.1 quantify this statement. The derived 'on' timescales are longer, *and* jet powers higher in the most massive hosts. In other words, higher radio luminosities imply more powerful, older sources rather than younger ones that have not yet suffered significant adiabatic, synchrotron and inverse-Compton losses. Fits to the radio-loud fraction broken up by stellar mass (as in Figure 10.1) then give the duration of the 'off' phase. It can be seen that massive galaxies not only spend a greater fraction of their time in the active state, but also longer in the 'on' phase and less time in the 'off' phase in absolute terms.

10.2.4 Implications for AGN triggering

What do these findings actually tell us? More specifically, do they shed any light on radio source triggering? We expect the radio activity to be limited by availability of cold gas. This will eventually be depleted due to AGN feedback, and hence jet production will cease. Since a higher rate of cooling and lower consumption rate will yield a longer lasting jet, we expect $t_{\text{on}} \propto \dot{M}_{\text{cold}} / Q_{\text{jet}}$. Following the arguments of Best *et al.* (2005b), the cooling rate is related to temperature and X-ray luminosity via $\dot{M}_{\text{cold}} \propto L_X / T$ and for isothermal gas $T \propto \sigma^2$. The correlation between X-ray and optical emission in luminous elliptical galaxies (Sullivan *et al.* 2001)

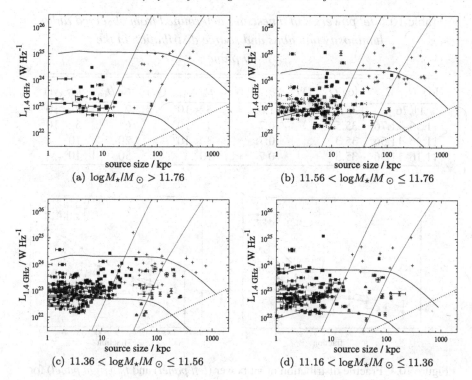

Figure 10.3 Distribution of radio-loud sources in the $P-D$ plane. Different symbols correspond to different types of sources. Crosses represent sources whose size is determined by inspection. Filled triangles are NVSS, and squares FIRST sources. Open symbols represent unresolved sources. The dashed line in the bottom right corner shows the limits on maximum detectable source size in the sample. $P-D$ tracks with jet power corresponding to mean $\pm\sigma$ (Table 10.1) are also shown as thick solid lines. Finally, dotted lines show the expected tracks corresponding to source ages of 10^7 and 10^8 years; smaller source sizes correspond to younger sources.

and the Faber–Jackson relation (Faber and Jackson 1976) give $L_X \propto L_{\mathrm{opt}}^4 \propto \sigma^8$, and thus $\dot{M}_{\mathrm{cold}} \propto \sigma^6$, or $\dot{M}_{\mathrm{cold}} \propto M_{\mathrm{BH}}^{1.5}$ on using $M_{\mathrm{BH}} \propto \sigma^4$ (Gebhhardt *et al.* 2000). Finally, using the $M_{\mathrm{BH}} \propto M_{\mathrm{bulge}}$ relation (Häring and Rix 2004), $\dot{M}_{\mathrm{cold}} \propto M_{\mathrm{bulge}}^{1.7}$. Figure 10.4 and Table 10.1 suggest that the observed jet power scales as $Q_{\mathrm{jet}} \propto M_{\star}^{0.6\pm0.2}$, and thus under the assumption of a constant spheroid-to-total mass ratio we expect $t_{\mathrm{on}} \propto M_{\star}^{1.1\pm0.5}$. This is consistent with the derived t_{on} values, where an increase in stellar mass by a factor of 4 results in an increase in t_{on} by a factor of about 5. In other words, the results are consistent with a picture in which the radio sources are mostly powered by the cooling of hot gas, and stay switched on until they run out of fuel. It is worth noting that this result is largely independent of the adopted density profile: if one were to let the central density ρ_{core} (Equation 10.1) scale with stellar mass, the derived jet powers would be largely independent of

Table 10.1 *Jet powers and timescales determined from observed radio
luminosity functions and source distribution in the
P–D plane*

$\log M_\star/M_\odot$	$\log \bar{Q}_{\text{jet}}$ (W)	$\sigma_{Q_{\text{jet}}}$ (W)	$t_{\text{on,median}}$ (yr)	$t_{\text{off}}/t_{\text{on}}$	t_{off} (yr)
> 11.76	35.6	1.0	5×10^6	2	2×10^7
11.56–11.76	35.5	0.7	4×10^6	6	5×10^7
11.36–11.56	35.3	0.7	3×10^6	20	10^8
11.16–11.36	35.3	0.7	10^6	50	10^8

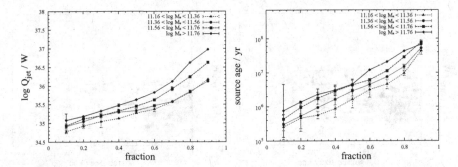

Figure 10.4 Fractile distribution of jet power (*left panel*) and t_{on} (*right panel*) for
the top four stellar mass bins.

mass, but the strong t_{on} dependence remains. In the absence of heating, the time
a galaxy spends in the quiescent phase should be $t_{\text{off}} \propto \dot{M}_{\text{cool}}^{-1} \propto M_\star^{-1.7\pm0.3}$. The
observed relation (Table 10.1) is nowhere near as steep, probably largely due to
the fact that massive galaxies host the most powerful radio sources. These affect
their environments on much larger spatial scales, which (due to the longer cooling
times) correspond to longer timescales, than their low mass counterparts. As a
result, massive galaxies spend longer in the quiescent phase than is expected from
simple cooling arguments.

10.3 Intermittent AGN feedback in galaxy formation

Some obvious questions to ask are: how does this mode of intermittent feedback
affect the formation of structure? Does it explain cosmic downsizing and the lack of
brightest galaxies predicted by hierarchical models by suppressing star formation
in massive galaxies at late times?

Growth of structure is typically traced by semi-analytic models. In these, evolution of dark matter halos is described by detailed numerical simulations, while
baryonic matter is followed via analytical prescriptions. Gas accreted onto the dark

matter halo cools and eventually forms stars, subject to a number of feedback mechanisms. These include heating from the ionising UV flux of the first generation of stars and quasars; supernovae feedback as massive stars collapse in on themselves; and AGN feedback in the more massive galaxies. The last channel in particular has received a lot of attention recently, with a number of groups attempting to address this mode of feedback (Granato *et al.* 2004; Bower *et al.* 2006; Cattaneo *et al.* 2006; Croton *et al.* 2006). In most cases the feedback prescriptions are either phenomenological or assume the AGN output power is coupled in a simple way to the surrounding gas. The discussion above (see also Chapters 12 and 14) suggests the radio source/environment interaction is not quite so simple, and in fact observations indicate the efficiency of such coupling is a function of galaxy mass (Best *et al.* 2007).

10.3.1 Galaxy formation framework

Shabala and Alexander (2009b) introduced a framework in which intermittent feedback from radio AGNs can be incorporated in galaxy formation models. Analytical fits of van den Bosch (2002) to the Millennium Simulation results were adopted to follow the growth of a dark matter halo of a given redshift-zero mass. This formulation implicitly includes mergers by averaging over them. Following the standard paradigm, gas is accreted supersonically along with the dark matter at the virial radius. This gas is initially shocked to the virial temperature and density, and can then cool and eventually form stars. Both the hot gas and dark matter were assumed to follow an NFW (Navarro *et al.* 1998) profile.

Star formation

Once the gas cools, it is transported to the galactic disk after a dynamical time. Stars form in the disk if there is a sufficient surface density of cold gas for the gas to become gravitationally unstable. This surface density threshold can be converted to a critical mass once the size of the disk is related to halo mass. The star formation rate is related to the amount of cold gas available and dynamical time of the disk.

Feedback in low mass galaxies

At the highest redshifts, feedback from the first generations of structure is important. The effect of this feedback is to limit the amount of accreted baryonic matter relative to the universal value, given by the ratio of baryonic to total matter density in the universe (Okamoto *et al.* 2008).

Stars more massive than about 8 $M_\odot$ end their lives as supernovae. These inject kinetic energy into the ICM, heating and ejecting gas out of the galaxy. Following a standard prescription in semi-analytic models, this feedback channel is mimicked by assuming a fraction of stars to be instantaneously recycled into cold disk gas. The amount of feedback energy is proportional to the recycled mass. Ejected gas returns to the galaxy within a few dynamical times. In the Shabala and Alexander model, this mode of feedback is parametrised by the efficiency parameter ϵ_{halo}.

Black hole growth and radio source feedback

Reionisation feedback is important at early cosmic times, while supernovae feedback is the dominant channel at late times in galaxies less massive than M^*. In the most massive hosts, AGN feedback becomes important. A fraction of cold disk gas will be accreted onto the black hole, with $\dot{M}_{BH} \propto M_{cold}$. Shabala and Alexander (2009b) parametrise the accretion by ϵ_{acc}.

Gas angular momentum is removed by viscosity as it spirals in towards the black hole, generating jets. Jet properties depend on the nature of the accretion flow. At high accretion rates (relative to the Eddington rate) the accretion solution is described by the standard Shakura–Sunyaev (Shakura and Sunyaev 1973) thin disk. The flow is optically thick, and produces a quasi-blackbody spectrum. This flow is radiatively efficient, as all the energy released through viscous dissipation can be radiated away, and is typically identified with the bright AGN quasar phase. At low accretion rates, the flow becomes optically thin and geometrically thick. Instead of being radiated away, thermal energy of inflowing gas is advected inward. The resultant advection dominated accretion flow (ADAF) is radiatively inefficient, and can produce powerful jets (up to three orders of magnitude more powerful than in the thin disk phase (Meier 2001)). This is identified with the optically quiescent radio phase, associated with radio AGN activity. Thus, a jet is generated when the accretion disk flow switches to an ADAF. This takes place when the instantaneous accretion rate onto the black hole falls below the critical value $\dot{m}_{crit} = \dot{M}_{BH}/\dot{M}_{Edd}$, and the resultant jet power is $Q_{jet} = 0.79\dot{M}_{BH}c^2$ (Shabala and Alexander 2009b). The jet inflates a radio cocoon, which expands outwards, shock heating the ambient gas (Figure 10.2). This picture provides a self-regulatory model of intermittent feedback. Low gas cooling rates generate weak jets (as $\dot{M}_{BH}$ is low), and these get disrupted by Kelvin–Helmholtz instabilities in the dense central regions of the galaxy. There is no significant heating, and the accretion rate increases to yield a more powerful jet, which can regulate the cooling. If the jet is too powerful, however, the bulk of the energy is deposited at large radii, and central cooling can proceed until the accretion rate exceeds $\dot{m}_{crit}\dot{M}_{Edd}$. At this point the accretion flow changes to a thin disk solution, and the jet terminates until the black hole grows massive enough for the accretion rate (in Eddington units) to fall below this threshold, whence the whole process can restart. Such a picture is consistent

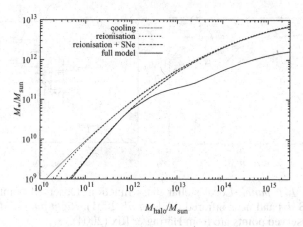

Figure 10.5 Stellar mass at the present epoch for various models.

with the above observational result that in local galaxies the presence of radio jets is determined by the amount of cold gas cooling out to fuel the central engine.

A key question is: at what $\dot{m}_{crit}$ does the accretion disk change its state? The transition from an ADAF to thin disk accretion is fairly well defined. Theoretical considerations (Narayan *et al.* 1998) and observations of galactic X-ray binaries (Körding *et al.* 2006) suggest $\dot{m}_{crit,up} \sim 0.1$. The reverse transition, corresponding to the onset of the radio jet, is somewhat more complicated. A thin disk flow will gradually develop a growing hole in its centre, where the accretion flow becomes optically thin and hence an ADAF is allowed. Shabala and Alexander (2009b) chose to treat the dimensionless accretion rate below which the thin disk solution changes to an ADAF as a free parameter $\dot{m}_{crit,down}$ in the range 0.01–0.1.

10.3.2 *Local observations*

There are three free parameters in the model: supernovae feedback is parametrised by ϵ_{halo}, accretion rate onto the black hole by ϵ_{acc}, and radio jet triggering occurs when $\dot{M}_{BH} < \dot{m}_{crit,down} \dot{M}_{Edd}$. All of these can be constrained by local observations.

Figure 10.5 shows the dependence of the redshift-zero stellar mass in a given halo on the various modes of feedback. Reionisation and supernovae feedback are important in halos less massive than $\sim 10^{12}$ $M_{\odot}$ and at high redshifts, while AGN feedback dominates the $z \sim 0$ signature in massive galaxies. Thus, supernovae feedback efficiency can be set by the low-mass end of the local stellar mass function. The predicted mass function comes from convolving the redshift-zero stellar masses of various halos (Figure 10.5) with the halo mass function derived from N-body simulations (Jenkins *et al.* 2001). Figure 10.6 shows that supernovae and reionisation feedback alone cannot explain the dearth of massive galaxies.

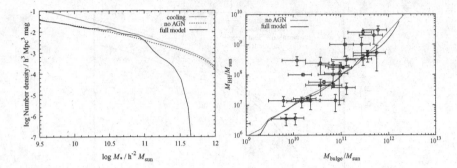

Figure 10.6 *Left panel*: Local stellar mass function. Observed points are from the 2dFGRS J-band near-infrared (Cole *et al.* 2001). *Right panel*: $M_{BH}-M_{bulge}$ relation. Observed points are from Häring & Rix (2004).

With supernovae feedback efficiency fixed and assuming a bulge-to-total stellar mass fraction, the low-mass end of the $M_{BH}-M_{bulge}$ relation (Häring and Rix 2004) can then be used to constrain the efficiency of black hole accretion, ϵ_{acc}. Finally, the parameter $\dot{m}_{crit,down}$ describing the onset of the radio phase is found by matching the local stellar mass function at the bright end. Figure 10.6 shows that the best-fit values for these parameters provide a good fit to observations. Importantly, the build-up of bulge and black hole mass remains 'in step' even when AGN feedback is included.

10.3.3 Cosmological evolution

The model reproduces observed properties of the local galaxy population. With all parameters constrained, past epoch predictions can be made. Figure 10.7 shows the evolution in the star formation rate density, found by convolving the model star formation histories with the halo mass function.

In the absence of AGN feedback, the model overpredicts star formation rate (SFR) density at the present epoch, consistent with the overabundance of massive galaxies in Figure 10.6. With AGNs included, the fit to SFR densities is good to $z \sim 1.5$–2.

Figure 10.8 shows the evolution of the stellar mass function to $z = 2.6$. As with the SFR density, the mass assembly history is reproduced extremely well out to $z \sim 1.5$, including the early formation of the most massive galaxies. At higher redshifts the model significantly underpredicts number counts of the most massive galaxies. It has recently been argued (Baugh *et al.* 2005) that a more top-heavy initial mass function (IMF) may be appropriate at high z. Physically this would be due to a change in the dominant star formation mechanism from merger-driven to *in situ* (in disks). Although the jury is still out on this issue, observational evidence

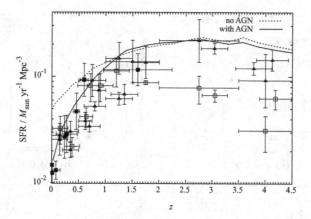

Figure 10.7 Evolution of the global SFR density. Observations are from the compilation of Hopkins (2004; filled points) and *J*-band observations of Cole *et al.* (2001; open squares).

based on metallicity distributions and Ly-break galaxy counts provides support for this idea. Merger-driven starbursts would produce a higher fraction of massive stars, resulting in lower mass-to-light ratios and thus making the derived stellar masses at high z overestimates. The $z = 2.0$ and $z = 2.6$ panels of Figure 10.8 show that a change to a top-heavy IMF (open symbols) gives good agreement between the model and observations, consistent with a change in the dominant star-formation mechanism at $z \sim 2$.

10.3.4 Cosmic downsizing

From the evolution of the stellar mass function (Figure 10.8) it is clear that the most massive galaxies are assembled first, and the less massive galaxies are ones undergoing star formation at the present epoch. The effect of AGN feedback on hierarchical structure formation can be examined by considering SFR densities as a function of stellar mass (Figure 10.9). Inclusion of feedback has little effect on low-mass galaxies, but suppresses the star formation heavily in the most massive galaxies since $z \sim 1.5$, resulting in 'cosmic downsizing'. The high-redshift in the most massive galaxies are somewhat overpredicted. This is most likely due to a vital missing ingredient in the model: powerful AGN feedback. Shock heating from radio sources is a feedback channel available to both Fanaroff–Riley (FR) type I and II sources. However, the less powerful FR-Is are disrupted by interaction with the ambient gas and are thus typically much smaller than the more powerful FR-IIs. As a result, they are much less efficient at transporting dense gas away from galaxy

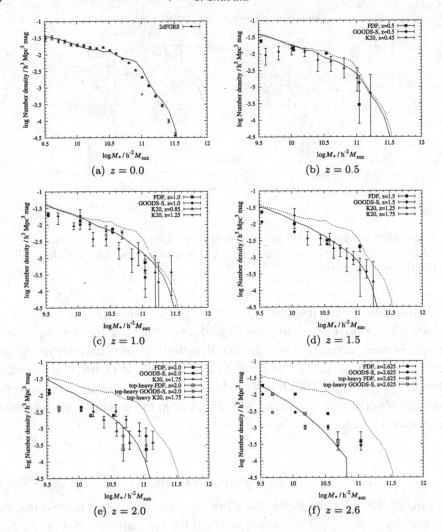

Figure 10.8 Evolution of the stellar mass function (solid lines). The corresponding quantity at $z = 0$ is shown as a dashed line for comparison. Observed data points are from 2dFGRS (Cole *et al.* 2001), GOODS-South and FORS Deep Field (Drory *et al.* 2005) and the K20 survey (Fontana *et al.* 2004). Closed symbols correspond to the usual diet Salpeter IMF (Bell *et al.* 2003) used in all preceding plots; open symbols are for a top-heavy Bottema IMF (Bottema 1997).

centres to large radii (where the cooling times are long). The absence of this 'gas uplifting' mode of feedback in the model is likely responsible for the disagreement with observations in Figure 10.9.

AGN feedback also naturally gives rise to the colour bimodality of galaxies. Figure 10.10 shows a sharp break in both the mass- and luminosity-weighted mean stellar ages, with massive galaxies hosting significantly older stellar

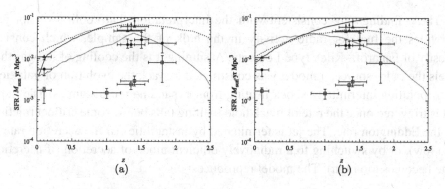

Figure 10.9 Star formation rate densities broken up by stellar mass: (a) with AGN feedback; (b) no AGN. All curves are for: $10^{9.26} < M_\star/M_\odot < 10^{10.46}$ (solid line, open triangles), $10^{10.46} < M_\star/M_\odot < 10^{11.06}$ (dashed line, closed circles), $10^{11.06} < M_\star/M_\odot < 10^{11.76}$ (dotted line, open squares). Data points for $z > 0.5$ are from Juneau *et al.* (2005) and from Brinchmann *et al.* (2004) for the local volume.

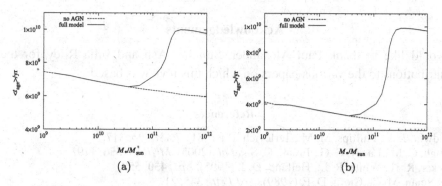

Figure 10.10 Mean stellar ages weighted by mass (a) and luminosity (b).

populations. Significantly, the location of the predicted break is consistent with a similar observed feature at $M_\star \sim 3 \times 10^{10} \, M_\odot$ in SDSS galaxies (Kauffmann *et al.* 2003).

10.4 Summary

Studies of local radio galaxies and their host properties show that massive galaxies host a higher fraction of radio sources. This has been interpreted as due to more frequent re-triggering. Fitting radio source models to derive source ages and active timescales suggests that massive galaxies:

- host the more powerful jets;
- radio sources are 'on' for longer;
- spend less time in the radio-quiet phase.

These findings are consistent with the interpretation that cooling of hot gas provides the fuel for radio activity in the bulk of the sample, which consists mostly of Fanaroff–Riley type I objects. Assuming it is the cooling of hot gas that fuels the radio source, a model was constructed to track the evolution of galaxies, incorporating intermittent shock heating from expanding radio sources. The radio jet is triggered once the rate of black hole fuelling falls below some critical fraction of the Eddington rate. The jet is terminated by instabilities (if the accretion rate is too low), or by switching to a radiatively efficient mode of accretion (if accretion rate becomes too high). The model reproduces:

- the local stellar mass function;
- evolution in stellar mass function and SFR density to $z \sim 1.5-2$, including early formation of massive galaxies;
- a top-heavy initial mass function, corresponding to a change in the dominant mode of star formation, is required to explain observations at $z > 2$.

Acknowledgements

I would like to thank Paul Alexander, Summer Ash and Julia Riley for their contributions to the various papers on which this review is based.

References

Baldwin, J. A., Phillips, M. M., Terlevich, R. (1981). *PASP*, **93**, 817.
Baugh, C. M., Lacey, C. G., Frenk, C. S., *et al.* (2005). *MNRAS*, **356**, 1191.
Becker, R. H., White, R. L., Helfand, D. J. (1995). *ApJ*, **450**, 559.
Begelman, M. C., Cioffi, D. F. (1989). *ApJ Lett.*, **345**, 21.
Bell, E. F., McIntosh, D. H., Katz, N., Weinberg, M. D. (2003). *ApJ Supp.*, **149**, 289.
Best, P. N., Kauffmann, G., Heckman, T. M., Ivezić, Ž. (2005a). *MNRAS*, **362**, 9.
Best, P. N., Kauffmann, G., Heckman, T. M., *et al.* (2005b). *MNRAS*, **362**, 25.
Best, P. N., von der Linden, A., Kauffmann, G., Heckman, T. M., Kaiser, C. R. (2007). *MNRAS*, **379**, 894.
Bottema, R. (1997). *A&A*, **328**, 517.
Bower, R. G., Benson, A. J., Malbon, R., *et al.* (2006). *MNRAS*, **370**, 645.
Brinchmann, J., Charlot, S., White, S. D. M., *et al.* (2004). *MNRAS*, **351**, 1151.
Burns, J. O. (1990). *AJ*, **99**, 14.
Cattaneo, A., Dekel, A., Devriendt, J., Guiderdoni, B., Blaizot, J. (2006). *MNRAS*, **370**, 1651.
Cole, S., Norberg, P., Baugh, C. M., *et al.* (2001). *MNRAS*, **326**, 255.
Condon, J. J., Cotton, W. D., Greisen, E. W., *et al.* (1998). *AJ*, **115**, 1693.
Cowie, L. L., Songaila, A., Hu, E. M., Cohen, J. G. (1996). *AJ*, **112**, 839.
Croton, D. J., Springel, V., White, S. D. M., *et al.* (2006). *MNRAS*, **365**, 11.
Drory, N., Salvato, M., Gabasch, A., *et al.* (2005). *ApJ Lett.*, **619**, 131.
Faber, S. M., Jackson, R. E. (1976). *ApJ*, **204**, 668.
Fabian, A. C., Sanders, J. S., Allen, S. W., *et al.* (2003). *MNRAS Lett.*, **344**, 43.
Fontana, A., Pozzetti, L., Donnarumma, I., *et al.* (2004). *A&A*, **424**, 23.

Forman, W., Nulsen, P., Heinz, S., *et al.* (2005). *ApJ*, **635**, 894.
Gebhardt, K., Bender, R., Bower, G., *et al.* (2000). *ApJ Lett.*, **539**, 13.
Granato, G. L., De Zotti, G., Silva, L., Bressan, A., Danese, L. (2004). *ApJ*, **600**, 580.
Häring, N., Rix, H.-W. (2004). *ApJ*, **604**, 89.
Hopkins, A. M. (2004). *ApJ*, **615**, 209.
Jenkins, A., Frenk, C. S., White, S. D. M., *et al.* (2001). *MNRAS*, **321**, 372.
Juneau, S., Glazebrook, K., Crampton, D., *et al.* (2005). *ApJ Lett.*, **619**, 135.
Kaiser, C. R., Alexander, P. (1997). *MNRAS*, **286**, 215.
Kaiser, C. R., Dennett-Thorpe, J., Alexander, P. (1997). *MNRAS*, **292**, 723.
Kauffmann, G., Heckman, T. M., White, S. D. M., *et al.* (2003). *MNRAS*, **341**, 54.
Körding, E. G., Jester, S., Fender, R. (2006). *MNRAS*, **372**, 1366.
Magorrian, J., Tremaine, S., Richstone, D., *et al.* (1998). *AJ*, **115**, 2285.
Meier, D. L. (2001). *ApJ Lett.*, **548**, 9.
Narayan, R., Mahadevan, R., Quataert, E. (1998). In *Theory of Black Hole Accretion Disks*, eds. M. A. Abramowicz, G. Bjornsson, J. E. Pringle (Cambridge: Cambridge University Press).
Navarro, J. F., Frenk, C. S., White, S. D. M. (1997). *ApJ*, **490**, 493
Okamoto, T., Gao, L., Theuns, T. (2008). *MNRAS*, **390**, 920.
O'Sullivan, E., Forbes, D. A., Ponman, T. J. (2001). *MNRAS*, **328**, 461.
Shabala, S. S., Alexander, P. (2009a). *MNRAS*, **392**, 1413
Shabala, S. S., Alexander, P. (2009b). *ApJ*, **699**, 525.
Shabala, S. S., Ash, S., Alexander, P., Riley, J. M. (2008). *MNRAS*, **388**, 625.
Shakura, N. I., Sunyaev, R. A. (1973). *A&A*, **24**, 337.
van den Bosch, F. C. (2002). *MNRAS*, **331**, 98.
York, D. G., Adelman, J., Anderson, Jr., J. E., *et al.* (2000). *AJ*, **120**, 1579.

11

Environment or outflows? New insight into the origin of narrow associated QSO absorbers

V. Wild

11.1 Introduction

AGN feedback is widely proposed as the solution to a number of otherwise difficult-to-explain problems in extra-galactic astrophysics. From an observational perspective, it is worth first dissecting the forms of 'feedback' that are under discussion, before embarking on any project to observe this potentially universal process. Figure 11.1 gives a short summary of the topic of feedback, which can broadly be split into two parts (column 2): heating of gas *in situ*, and outflows that remove matter from the host galaxy. Both processes may, or may not, be associated with jets, so jets have been placed separately. While outflows are assumed to predominantly affect the nuclear region and possibly the ISM of the host galaxy, *in-situ* heating of the gas must occur on very large scales within the IGM (column 3). The final column presents a selection of observed or yet-to-be-observed consequences of the physical mechanisms: the list is not meant to be exhaustive, but simply presents the range of the observations with which we must deal. While there is little argument that some aspects of AGN feedback have been directly detected, conclusive evidence for routine quenching of star formation and removal of the interstellar medium of the QSO host galaxy remains elusive.

Of particular relevance to this contribution are the narrow absorption line systems (NALs), which appear in every box on the right-hand side of Figure 11.1, and are arguably one of the best candidates for *directly* detecting 'ubiquitous' QSO feedback. These absorption lines, generally detected in the rest-frame ultraviolet due to the convenient gathering of several strong transitions, are caused by clouds of ionised and/or neutral gas that intervene between a strong light source and the observer. This material need not be dense to cause significant absorption of the traversing light: the detection of MgII absorption generally implies hydrogen

AGN Feedback in Galaxy Formation, eds. V. Antonuccio-Delogu and J. Silk. Published by Cambridge University Press. © Cambridge University Press 2011.

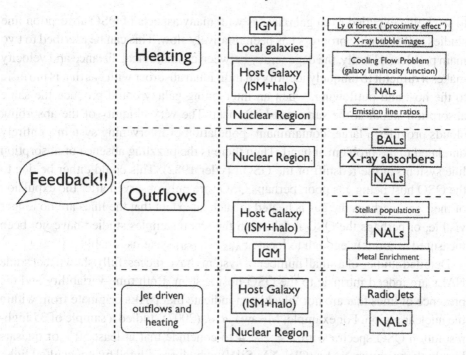

Figure 11.1 What do we mean by AGN feedback?

column densities of a few 10^{17} atoms/cm^2, for CIV this number is two orders of magnitude smaller at $\sim 10^{15}$ atoms/cm^2. In contrast to the well-studied broad absorption line systems (BALs), NALs are thought in general to arise from the interstellar medium and surrounding halo gas of ordinary galaxies. Together with the convenient placing of their resonance transition lines in the observed-frame optical at redshifts of interest for QSO feedback, metal NALs can be a very powerful tool for probing the physical state and position/velocity of gas both within and external to the host galaxies.

In these proceedings I will review some of the studies using NALs to look for direct evidence of QSO feedback, from detailed studies of a few objects through to statistical studies using the largest databases of absorbers available to us today. I will present new results on the distribution of line-of-sight velocity offsets between MgII absorbers and their background QSOs, which reveal a high-velocity population similar to that observed recently for CIV.

11.2 Using ultraviolet NALs to reveal QSO feedback

Although there is an abundance of data on NALs, covering most of the age of the universe, and providing the potential to trace the very gas clouds we hope to see being expelled from galaxies, as with many aspects of QSO absorption line studies real scientific progress has been relatively slow. This can be ascribed to two main problems. Firstly, the degeneracy between cosmological distance and velocity makes it difficult to uniquely identify an individual absorber with gas that is intrinsic to the host and outflowing, when an intervening galaxy could produce the same absorption signal at the same redshift offset. The very ubiquity of the absorbing clouds provides a large 'contaminant' population of intervening systems, entirely unrelated to the problem at hand. Then there is the puzzling absence of absorption line systems at the redshift of the QSO (Tytler 1982). This could either be due to the QSO host being gas poor, perhaps QSOs are only observed after the expulsion of their gas, perhaps the gas is heated to such an extent that the lines are no longer visible, or perhaps the QSO redshifts in the small samples studied have not been measured accurately enough to locate $z_{ABS} \sim z_{QSO}$ systems reliably.

Detailed studies of a small number of systems have successfully shown that some NALs are indeed intrinsic to the QSO-host system. Both time-variability and the presence of lines that are not 'black' can indicate that NALs originate from within the nuclear region. For example, Misawa et al. (2007) studied a sample of 37 high-resolution QSO spectra with $2 < z < 4$ to conclude that at least 50% of quasars host high ionisation NALs (CIV, NV, SiIV), which are diluted by unocculted light, and thus lie close to the central engine. Detailed observations of multiple transitions have resulted in mass loss rates and precise distances for a handful of objects at both high and low redshift (e.g. Crenshaw et al. 2007; Rix et al. 2007). However, with such small samples, and especially at high redshift where imaging is difficult in front of a bright background QSO, the question always remains as to whether an absorption system at a few kpc is simply the sign of an intervening galaxy. Detailed analyses of NALs in QSO sightlines that pass close to foreground QSOs also allow us to probe the effect of the QSO on gas that does not lie directly within the 'firing range' of the QSO (the 'transverse proximity effect'). Recent results remain inconclusive: Bowen et al. (2006) find no evidence for a reduction in strong MgII systems, whereas Gonçalves et al. (2008) find a significant change in ionisation state of gas on scales of 1 Mpc. Hennawi and Prochaska (2007) detect an isotropy in the distribution of 17 Lyman-limit systems around QSOs, suggesting that the line-of-sight systems may be photoevaporated. Such detailed analyses of small numbers of systems have been complemented by the statistical analyses of large samples of NALs. Until recently the question primarily revolved around the presence, or absence, of an excess of absorbers close to QSOs (so called 'associated' systems, with velocities from below a few hundred to a few thousand km/s depending on the study). With large samples, an excess of absorbers at $z_{ABS} \sim z_{QSO}$ has now been clearly detected (e.g. Richards 2001; Vestergaard 2003). Vanden Berk et al. (2008) compared the properties of associated absorbers to those at larger redshift separations, finding that they are dustier and have higher ionisation states. But the ambiguity remains as to whether the population arises from neighbouring galaxies or from gas associated with the QSO, its host galaxy and its halo.

11.3 The line-of-sight distribution of NALs in front of QSOs

The Sloan Digital Sky Survey (SDSS) has led to an enormous increase in the quantity of data available on QSO absorption line systems. While the spectra are not of particularly high resolution or signal-to-noise ratio (SNR), preventing detailed analyses of individual systems, the shear numbers of objects allow statistical studies that were previously impossible. Here we present the statistical analysis of associated NALs, through a new analysis of MgII absorption line systems based on a catalogue of nearly 20 000 systems culled from the sixth data release (DR6) of the SDSS survey. The catalogue was constructed using a matched-filter detection algorithm as described in Wild *et al.* (2006). For reasons of catalogue completeness, the NALs are restricted to rest-frame equivalent width $W_{\lambda2796} > 0.5$ and the QSO spectra searched are required to have per-pixel-SNR > 8. For the analysis of $z_{ABS} \sim z_{QSO}$ absorption systems, accurate redshifts for the QSOs are crucial (Nestor *et al.* 2008), a difficult problem due to the substantial broadening of the emission lines in QSOs. The results presented here rely upon new QSO redshifts using a combination of available narrow emission lines and new cross-correlation templates (Hewett & Wild 2010).

In Figure 11.2 we present the distribution of velocity offsets between the MgII absorption line systems and their background QSOs:

$$\beta = \frac{R^2 - 1}{R^2 + 1}, \quad \text{where} \quad R = \frac{1 + z_{QSO}}{1 + z_{ABS}}. \tag{11.1}$$

For the first time for MgII, we can clearly identify three populations:

- At large velocities, $\beta > 0.02$, the constant number density is consistent with an intervening population of absorbers caused by galaxies and gas clouds that are not physically associated with the QSO.
- A clear spike in the numbers is seen at $\beta = 0$, consistent with a Gaussian distribution with mean of approximately zero and width of a few hundred km/s. Whether these NALs primarily originate in galaxies clustered around the QSO or in the QSO host galaxy is the question we must address.
- Finally, there is a very clear extended excess of absorbers out to velocities $\beta < 0.02$ or $v < 6000$ km/s, a feature previously seen clearly in CIV (Richards *et al.* 1999; Nestor *et al.* 2008; Wild *et al.* 2008), but only hinted at before for MgII (Wild *et al.* 2008).

V. Wild

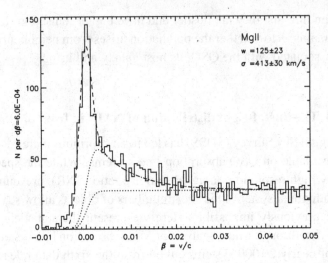

Figure 11.2 The distribution of line-of-sight velocity offsets between MgII NALs and their background QSOs (histogram). Overplotted as a dashed line is a toy model composed of three populations: low-velocity absorbers; high-velocity absorbers; and intervening absorbers with a constant space density (see text). Dotted lines indicate the individual contributions from the latter two components. The model is convolved with a Gaussian kernel to account for redshift errors and/or peculiar velocities.

The distribution is well described by a low-velocity component (delta function) centred approximately on zero, an exponentially distributed high-velocity component of width w at $\beta > 0$ (upper dotted line), both superposed on a constant background intervening population (B, lower dotted line), and convolved with a Gaussian kernel to account for redshift errors and/or peculiar velocities:

$$N_{ABS} = \left(A_1 \delta(\beta - \mu) + \left[A_2 \exp(w\beta) + B \right]_{\beta > 0} \right) * G(\sigma), \tag{11.2}$$

where we find $\sigma = 413 \pm 30\,\text{km/s}$ and $w = 125 \pm 23$. The full fit is shown as a dashed line in Figure 11.2. Unfortunately, the detection of an excess of NALs at the redshift of the QSO is not unambiguous evidence for NALs in the host galaxies of the QSOs. As we show in the next section, galaxy clustering can lead to a signal that is difficult to distinguish with current data.

11.4 The 3D distribution of NALs around QSOs

With the size of catalogues now available from the SDSS, it is possible to measure directly the 3D clustering of absorbers around QSOs using a cross-correlation style analysis. With the clustering amplitude in hand, we can then estimate the excess number of absorbers expected along the line-of-sight to the QSOs. The

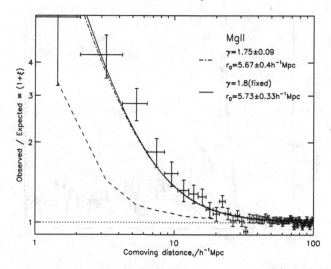

Figure 11.3 The 3D clustering of MgII NALs around QSOs as a function of comoving separation. The dashed line is the 68 per cent detection threshold given the number of sightlines and number density of absorbers. The dash-dot and solid lines are best-fit power laws with parameters given in the top right.

method used to measure the 3D clustering is presented in detail in Wild *et al.* (2008). To summarise, we count the number of observed QSO–MgII pairs (N_{obs}) as a function of comoving separation (r) and compare this to the number of pairs that would be expected (N_{exp}) for a constant background distribution of absorbers without clustering. To avoid contamination from NALs that might be associated with outflowing gas from a QSO host, we restrict the NAL sample to those with $z_{ABS} < z_{QSO} - 0.1$. In Figure 11.3 we present the new results using the DR6 MgII catalogue with improved QSO redshifts. The dash-dot line is a powerlaw fit of the form:

$$\xi(r) = \frac{N_{obs}}{N_{exp}} - 1 = (r/r_0)^{-\gamma}, \qquad (11.3)$$

where r_0 is the correlation scale length, which we measure to be $5.67 \pm 0.4 h^{-1}\text{Mpc}$, with a power law index of $\gamma = 1.74 \pm 0.09$, or $5.73 \pm 0.3 h^{-1}\text{Mpc}$ at fixed $\gamma = 1.8$. This correlation length is similar to that measured for bright galaxies at similar redshifts. There is evidence, at the level of around 3σ, for a flattening in the MgII–QSO clustering on small scales ($<5 h^{-1}$ Mpc). This may be caused by QSO redshift errors or absorber peculiar velocities, which can have a significant effect at small absorber–QSO separations. It may also indicate the presence of a transverse 'proximity effect', where the QSO ionises the gas in its surrounding halo (but see Bowen *et al.* (2006)). Clearly there is scope for further investigation of this feature in the future. Discarding the central bin from our power law fit increases

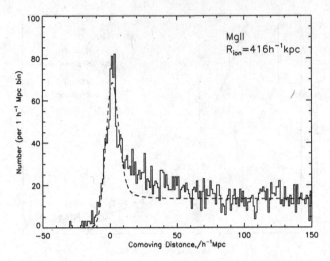

Figure 11.4 Distribution of line-of-sight QSO–absorber separation as a function of comoving distance (where line-of-sight redshift separation has been converted into comoving distance in the usual way). The dashed line shows the predicted distribution of MgII absorbers from clustering, assuming the QSO ionises (i.e. removes) all absorbers to a proper distance of $130h^{-1}$ kpc.

both the measured correlation length and power law index by about 3σ, leading to a larger predicted clustering signal that only enhances the qualitative conclusions drawn from this study.

11.5 The clustering contribution to the line-of-sight excess

In Figure 11.4 we convert the distribution of line-of-sight QSO–absorber redshift separations into comoving distance units, to allow direct comparison with the 3D-clustering results of the previous section. One free parameter is then required for our toy-model: the distance below which physical processes internal to the host galaxy dominate the distribution of absorbers, rather than the clustering of galaxies in the QSO neighbourhood. As we shall see, the precise distribution of MgII absorbers around the host galaxy of the QSOs is irrelevant to our results, due to the significant deficit of absorbers detected along the line-of-sight. We therefore define a simple 'ionisation radius' (R_{ion}) internal to which the number of absorbers is zero. Finally, the model is convolved with a Gaussian of width equivalent to $\sigma = 413$ km/s at the median redshift of absorbers that lie within $\pm 10h^{-1}$ Mpc of their background QSO ($z = 1.3$), i.e. to match the measured width of the distribution in velocity space (Figure 11.2). The dashed line in Figure 11.4 shows the predicted line-of-sight distribution of absorbers in front of QSOs, from galaxy clustering alone and with an ionisation radius $\sim 420h^{-1}$ kpc (comoving units), or $180h^{-1}$ kpc (proper units at the median redshift of the sample). We note that this value for the ionisation radius is slightly lower than that given in Wild *et al.* (2008), likely resulting from the completely independent method used to select the absorbers. Typical MgII halos around galaxies can extend to $\sim 40h^{-1}$ kpc (proper) with almost unity covering fractions (Steidel 1993). Beyond this distance MgII halos are thought to be patchy, and can extend to distances of $70h^{-1}$ kpc (Zibetti *et al.* 2007; Kacprzak *et al.* 2008). Our result shows that the MgII ion, with an ionisation potential of 15.03 eV, is destroyed in clouds that lie at even greater distances from QSOs and thus far into the IGM. The spike of absorbers below $\beta < 0.002$, $v < 600$ km/s, or $R < 10h^{-1}$ Mpc, is entirely consistent with galaxy clustering from galaxies that lie beyond the ionisation zone of the QSO. However, we cannot rule out that R_{ion} is indeed even larger and the low-velocity absorbers are caused by denser, self-shielded clouds remaining within the ionisation zone, perhaps even intrinsic to the QSO host itself. Clearly the high-velocity tail is, however, caused by a process internal to the QSO itself, and with velocities as high as $0.02c$ (~ 6000 km/s) these outflows must be driven by the central AGN engine, rather than any accompanying starburst (Tremonti *et al.* 2007). The very existence of these absorbers is puzzling, given the clear ability of the QSO radiation to destroy all normal MgII clouds out to very large distances. Their existence also leads us to question the conclusion that the low-velocity absorbers are primarily due to galaxy clustering. Are they the remnants of the densest ISM clouds yet to be destroyed? Are they unrelated to ordinary MgII ISM clouds, and instead created in the turbulence of outflowing gas? Their distance from the nuclear source remains to be determined. If they are external to the nuclear region, then they are surely evidence for the expulsion of (cold) gas from the galaxy ISM. If they are internal to the nuclear region, such low ionisation gas with narrow velocity widths can constrain models for the inner regions of QSOs (Elvis 2000).

11.6 Radio loud vs. radio quiet

One further statistical investigation may lead to significant insight into the origin of the MgII absorbers within 6000 km/s from the QSO. It has been known for some time that QSOs with different radio properties (loud/quiet, flat/steep spectrum) have different fractions of absorption line systems, strongly suggesting a non-intervening origin at least for a subset (Aldcroft *et al.* 1994; Richards 2001). More recently, detailed studies of nearby radio galaxies have revealed outflowing neutral hydrogen in 21 cm absorption against the background radio source (Morganti *et al.* 2005). In Figure 11.5 we present the velocity separation of radio loud QSOs (RLQSOs, $L_{\text{FIRST}} > 10^{25}$ W/Hz) compared to a sample of radio quiet QSOs

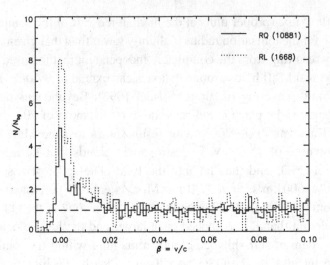

Figure 11.5 The distribution of line-of-sight velocity offsets between MgII NALs and the background QSOs, separated by the radio luminosity at 10^{25} W/Hz of the QSOs. The RQQSO sample has been selected to match the RLQSO sample in optical luminosity.

(RQQSOs) matched in optical luminosity to the RLQSOs. We can clearly see that RLQSOs show a larger excess of low-velocity MgII absorbers than RQQSOs with high significance. Within $-0.002 < \beta < 0.002$, RLQSOs have an excess of 6.2 absorbers over the background level, compared to 3.6 for RQQSOs. At high velocities, RLQSOs also seem to show a small increase in MgII NALs: for $0.002 < \beta < 0.02$, RLQSOs have an excess of 2.1 compared to 1.6 for RQQSOs. The question remains as to whether RLQSOs are more strongly clustered than RQQSOs. If so, then the excess low-velocity MgII absorbers seen in RLQSOs may result solely from them living in higher density neighbourhoods. Unfortunately, the SDSS DR6 catalogue is still not quite large enough to answer this question using the method presented above.

11.7 Conclusions

A number of recent studies have found that NALs intrinsic to the QSO host can be found in at least 50% of QSO spectra, and in most cases they are found to be outflowing. Simply due to observational limitations the position of these confirmed cases is, however, close to the central nucleus. In the few cases where larger distances can be determined from detailed line analyses, it is usually impossible to rule out the presence of an intervening galaxy.

Through the enormous statistical power of the SDSS, we can now determine precisely the contribution of galaxy clustering to QSO–absorber line-of-sight distributions. This leads to the following conclusions:

- QSOs heat the gas to considerable distances along their line-of-sight, with relatively low ionisation MgII ions ionised to several hundred kpc (comoving) into the IGM.
- Within 600 km/s there is an excess of NALs, however this excess is most simply explained by ordinary absorption clouds in and around galaxies that lie outside of the ionising influence of the QSO.
- A subset of absorbers out to velocities of 6000 km/s (MgII) or 12 000 km/s (CIV) cannot be explained by intervening galaxies. Their velocity distribution is well fit by a declining exponential (but see Nestor *et al.* (2008)), and their high maximum velocities indicate an origin close to the central engine.
- There is a significant excess of low-velocity NALs in RLQSOs, compared to RQQSOs. This excess may also extend into the high-velocity systems. Unfortunately, the statistics are not quite good enough to rule out the possibility that RLQSOs simply live in denser environments.

The heating effect of a QSO on its host galaxy, and likewise on all nearby galaxies, is unmistakable. However, the existence of the high-velocity systems, which we would naively expect not to exist in the intense radiation field of the QSO, leaves a narrow window of doubt as to the true origin of the low-velocity systems. Allowing the ionisation radius to increase, thus removing more intervening clouds, would allow some, if not all, low-velocity systems to arise from gas associated with the QSO, its host galaxy and its halo. There is certainly more work to be done before we can definitively claim the origin of low-velocity NALs to be intervening galaxies.

Acknowledgements

I would like to thank all of the team that helped me complete the first stage of this work, Paul Hewett for the new catalogue and QSO redshifts, the conference participants for their enthusiastic discussions and the conference organisers for putting together such an interesting programme.

Funding for the SDSS and SDSS-II has been provided by the Alfred P. Sloan Foundation, the Participating Institutions, the National Science Foundation, the U.S. Department of Energy, the National Aeronautics and Space Administration, the Japanese Monbukagakusho, the Max Planck Society, and the Higher Education Funding Council for England. The SDSS website is www.sdss.org.

References

Aldcroft, T. L., J. Bechtold, and M. Elvis. (1994). MgII absorption in a sample of 56 steep-spectrum quasars. *ApJS*, **93**, 1–46
Bowen, D. V., J. F. Hennawi, B. Ménard, *et al.* (2006). QSO absorption lines from QSOs. *ApJ*, **645**, L105–L108
Crenshaw, D. M. and S. B. Kraemer. (2007). Mass outflow from the nucleus of the Seyfert 1 galaxy NGC 4151. *ApJ*, **659**, 250–256
Elvis, M. (2000). A structure for quasars. *ApJ*, **545**, 63–76

Gonçalves, T. S., C. C. Steidel, and M. Pettini. (2008). Detection of the transverse proximity effect: Radiative feedback from bright QSOs. *ApJ*, **676**, 816–835

Hennawi, J. F. and J. X. Prochaska. (2007). Quasars probing quasars. II. The anisotropic clustering of optically thick absorbers around quasars. *ApJ*, **655**, 735–748

Hewett, P. C. and V. Wild. (2010). Improved redshifts for SDSS quasar spectra. *MNRAS*

Kacprzak, G. G., C. W. Churchill, C. C. Steidel, and M. T. Murphy. (2008). Halo gas cross sections and covering fractions of MgII absorption selected galaxies. *AJ*, **135**, 922–927

Misawa, T., J. C. Charlton, M. Eracleous, *et al.* (2007). A census of intrinsic narrow absorption lines in the spectra of quasars at $z = 2 - 4$. *ApJS*, **171**, 1

Morganti, R., C. N. Tadhunter, and T. A. Oosterloo. (2005). Fast neutral outflows in powerful radio galaxies: a major source of feedback in massive galaxies. *A&A*, **444**, L9–L13

Nestor, D., F. Hamann, and P. R. Hidalgo. (2008). The quasar-frame velocity distribution of narrow CIV absorbers. *MNRAS*, **386**, 2055–2064

Richards, G. T. (2001). Intrinsic absorption in radio-selected quasars. *ApJS*, **133**, 53–75

Richards, G. T., D. G. York, B. Yanny, *et al.* (1999). Determining the fraction of intrinsic CIV absorption in quasi-stellar object absorption-line systems. *ApJ*, **513**, 576–591

Rix, S. A., M. Pettini, C. C. Steidel, *et al.* (2007). The sightline to Q2343-BX415: Clues to galaxy formation in a quasar environment. *ApJ*, **670**, 15

Steidel, C. C. (1993). The properties of absorption-line selected high-redshift galaxies, in *The Environment and Evolution of Galaxies*, Proceedings of the 3rd Tetons summer school

Tremonti, C. A., J. Moustakas, and A. M. Diamond-Stanic. (2007). The discovery of 1000 km/s outflows in massive poststarburst galaxies at $z = 0.6$. *ApJ*, **663**, L77–L80

Tytler, D. (1982). QSO Lyman limit absorption. *Nature*, **298**, 427

Vanden Berk, D., P. Khare, D. G. York, *et al.* (2008). Average properties of a large sample of $z_{abs} \sim z_{em}$ associated MgII absorption line systems. *ApJ*, **679**, 239–259

Vestergaard, M. (2003). Occurrence and global properties of narrow CIVλ1549 absorption lines in moderate-redshift quasars. *ApJ*, **599**, 116–139

Wild, V., P. C. Hewett, and M. Pettini. (2006). Selecting damped Lyman α systems through CaII absorption – I. Dust depletions and reddening at $z \sim 1$. *MNRAS*, **367**, 211–230

Wild, V., G. Kauffmann, S. White, *et al.* (2008). Narrow associated QSO absorbers: clustering, outflows and the line-of-sight proximity effect. *MNRAS*, **388**, 227–241

Zibetti, S., B. Ménard, D. B. Nestor, *et al.* (2007). Optical properties and spatial distribution of MgII absorbers from SDSS image stacking. *ApJ*, **658**, 161–184

Part IV

Models and numerical simulations:
methods and results

12

Physical models of AGN feedback

V. Antonuccio-Delogu, J. Silk, C. Tortora, S. Kaviraj,
N. Napolitano & A. D. Romeo

12.1 Introduction

The idea that AGN activity can detectably influence the evolution of stellar populations in galaxies was advanced about 10 years ago (Silk and Rees 1998). This
feedback can either manifest itself in the form of episodes of induced star formation, as originally suggested by Silk and Rees, or one could also imagine that the
impressive release of energy from the AGN's jet to the interstellar medium (ISM)
of its host galaxy could inhibit stellar formation. The first form is called *positive*
feedback, while the second is often termed *negative* feedback. Both forms of feedback have durations of the order of a few times 10^7 years, i.e. the timescale of the
AGN's duty cycle. This is a very short timescale in terms of galaxy evolution: thus
the detection of negative feedback becomes possible by inspecting the statistical
properties of colour–colour and colour–magnitude diagrams in some bands that
are sensitive to recent star formation episodes. Only recently, with the massive
exploitation of data from large-scale surveys such as the Sloan Digital Sky Survey,
has it become possible to obtain galaxy samples large enough to check these effects
(see for instance the contribution by Silverman *et al.* in this volume, Chapter 4).

Only more recently, however, have simulations of the jet–ISM interaction been
attempted (see for instance the contributions by Bicknell *et al.* and Krause and
Gaibler in this volume, Chapters 14 and 16). Our recent efforts have been directed
at making more *realistic* simulations of this interaction, taking into account the
multiphase nature of the ISM. Our main task was to study in detail the evolution
of stellar formation within cold ISM clouds, when they are affected by an AGN.
This influence can take place in different ways: some clouds lie along the path of
the jet, others are embedded within the *cocoon* created by the jet when it sweeps
through the ISM of the host galaxy, and finally other clouds lie outside the region

AGN Feedback in Galaxy Formation, eds. V. Antonuccio-Delogu and J. Silk. Published by Cambridge
University Press. © Cambridge University Press 2011.

occupied by the cocoon at its maximum expansion. As we will see, even when using *adaptive mesh refinement* (AMR) numerical methods, stellar formation takes place on short temporal and spatial (*subgrid*) scales. Thus, we have designed the simulation setup with the purpose of following the thermodynamic evolution of the clouds with the highest possible resolution, and use these data as an input to model stellar formation.

In a first set of simulations, we studied in detail the evolution of a *single cloud*, lying in a region not far from the path of the jet: the results of this set of simulations are described in Section 12.2. Then we performed a more realistic simulation, where we included a set of 300 clouds, representing the cold component of the ISM, and we propagated a jet into this system. We then applied a model of stellar formation, and deduced the evolution of the *global* star formation in this system, and how it is affected by the jet. As we show in Section 12.3, after an initial transient phase of compression of the clouds from the bow shock of the cocoon, during which stellar formation is increased (*positive feedback*), the high temperatures reached within the cocoon act to significantly decrease the extension of the star-forming regions within the cold clouds, thus inducing a rapid decrease of star formation (*negative feedback*).

We then generalize this simulation by building a model of jet-modified star formation rate (SFR), which we apply to deduce the extension of the green valley and red cloud systems of galaxies affected by AGN feedback. Most galaxies from a sample recently studied by Kaviraj *et al.* (2007) lie within the region traced by the model, thus suggesting that their evolution could be the result of a fast quenching of the SF, possibly induced by the propagation of the jet.

12.2 Simulating jet propagation in a two-phase ISM

12.2.1 Numerical method

The simulations described here have been performed using FLASH v.2.5 (Fryxell *et al.* 2000), a parallel adaptive mesh refinement code, which implements a second-order, shock-capturing PPM solver. FLASH's modular structure allows the inclusion of physical effects such as external heating, radiative cooling and thermal conduction (among others). The most interesting aspect of using an AMR code such as FLASH for this problem lies in the considerable freedom permitted in the specification of the refinement criteria, which can be customized to reach very high spatial and temporal resolutions in selected regions. In particular, we have chosen refinement conditions that allow us to resolve in detail the clouds and their evolution. The main purpose of our work has been to study the interaction of a relativistic jet and the cocoon that it generates with pre-existing clouds, and how it can affect star formation within the clouds. Ideally, a full 3D simulation should have a spatial resolution high enough to resolve both the turbulent motions within the cocoon and the thermodynamic structure of the cloud, until the end of the simulation. The former requirement is important when one realizes that, in addition to the direct interaction with the jet, the cloud is also significantly affected by the random interactions with the turbulent eddies present within the cocoon. The computational requirements imposed by this task, however, are prohibitive: the smallest turbulent cells to be resolved should have a size comparable to that of the cloud ($10h^{-1}$ pc here), and the computational box is between 2 and 4×10^2 times larger. For this reason, we have restricted ourselves to 2D simulations, where this resolution can be easily reached.

We have included radiative cooling, described by a standard cooling function with half solar metallicity (Sutherland and Dopita 1993), extended to high temperatures ($T > 10^7$ K, see Appendix B of Antonuccio-Delogu and Silk (2008)). Gravity is included, while thermal conduction and magnetic fields are not. The former could possibly be relevant for the evolution of the cocoon, although on timescales longer than those considered here (Krause and Alexander 2007). Regarding magnetic fields, the evidence is that, if present within the diffuse intergalactic medium (IGM), their magnitude is no larger than a few microgauss, thus making the IGM a high-β plasma, and the magnetic field would then not significantly affect the global dynamical evolution.

In the simulations, the jet is modelled as a one-component fluid, with a density ρ_j, which is a fixed ratio ϵ_j of the environmental density ρ_{env}. In order to suppress the growth of numerical instabilities at the jet/IGM injection interface, we adopt a steep but continuous and differentiable transverse velocity and density profile previously adopted in simulations of jet propagation (Bodo *et al.* 1994; Perucho *et al.* 2004, 2005):

$$v_{x,j} = \frac{V_j}{\cosh\left\{(y - y_j)^{\alpha_j}\right\}}, \tag{12.1}$$

$$n_j = n_{env} - \frac{(n_{env} - n_j)}{\cosh\left\{(y - y_j)^{\alpha_j}\right\}}, \tag{12.2}$$

where $\alpha_j = 10$ is an exponent that determines the steepness of the injection profile and n_j, n_{env} denote the jet and environment number densities. This initial profile is highly sheared, and peaks around y_j, with most of the thrust $n_j v_j^2$ concentrated around the centre of the profile. The presence of a highly sheared injection profile forces the code to refine the grid structure, populating the injection region with subgrid meshes, thus preventing the formation of numerical contact instabilities.

Table 12.1 *Parameters of the simulation runs*

Run	L_{box} (h^{-1} kpc)	n_{env} (cm^{-3})	n_j/n_{env}	W_j (erg/s)
H0	4	1	0.02	4×10^{40}
H1	4	1	0.02	8.61×10^{41}
H2	8	1	0.01	1.34×10^{44}
H3	4	1	0.02	10^{45}
H4	4	1	0.02	10^{46}

Columns from left to right are as follows: run label, simulation box size, background density (runs H1–3) or central density (runs NFW), jet/background density ratio, jet power.

In all our runs, the injection point of the jet is chosen to lie at the midpoint of the left boundary.

12.2.2 Simulations

Our simulations are characterized by seven parameters. Two of these characterize the diffuse ISM: n_{env}, T_{env}, two describe the the cloud: r_{cl}, M_{cl}, and finally three characterize the jet: n_j, y_j, V_j. As the SFR is mostly determined by the strength of the shock within the cloud, as previous models seem to suggest (Klein *et al.* 1994), we have decided to mostly vary the jet's input power $P_j = 0.5A_j m_H n_j V_j^3$ which, together with the density contrast, determines the timescale of the cloud's disruption. We keep most of the other parameters fixed: ISM density and temperature at 1 e$^-$ cm^{-3} and 10^7 K, respectively, typical of the hot ISM in the central parts of a massive elliptical galaxy at high redshift ($z \approx 1$). Only in one run have we increased the size of the box, in order to check that the main features of the evolution do not depend on boundary conditions. All runs except one have been performed on a relatively small box ($4h^{-1}$ kpc), while in run H2 we used a box twice as large.

In Table 12.1 we summarize the main parameters of the different runs. The ratio between jet and environment density, n_j/n_{env}, is fixed to 2×10^{-2} in runs using small boxes, and decreased to a slightly lower value (10^{-2}) in run H2. The injection region has a width of $100h^{-1}$ pc (small box) and twice that width in run H2. The parameter V_j in Equation 12.1 is computed once ρ_j, d_j and P_j have been assigned. Finally, we chose *open* boundary conditions, so the gas is free to flow out of the simulation box. The implication of this is that gas is not allowed to re-enter the box, so circulation motions on scales larger than the simulation box cannot be reproduced.

Table 12.2 *Cloud parameters*

L_{box} (h^{-1} kpc)	x	y	r_{cl}	M_{cl}
4	0.6	1.92	10	1.65×10^8
8	1.2	3.84	20	1.30×10^9

All distances are expressed in h^{-1} kpc.
From left to right, columns are as follows: simulation box size, x and y coordinates, cloud radius and mass, the latter in $M_\odot$.

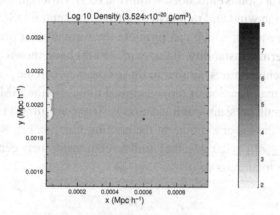

Figure 12.1 Initial configuration for the runs. The plot is a magnification of the actual simulation box, showing a small region of the larger simulation box. The small, dense cloud lying along the path of the jet, is placed at (0.6, 1.92) (in h^{-1} kpc).

12.2.3 *Initial configuration*

We place a small, dense cloud at a position slightly offset from the jet's propagation direction (see Figure 12.1 and Table 12.2). In order to keep the same spatial resolution, the radius of the cloud is doubled in run H2, where the simulation box is larger: consequently also the initial mass of the cloud is larger. We have chosen to study the effect of the jet on a cloud located very near to its path, because this allows us to check to what extent star formation is affected by the jet under the most extreme conditions. In Section 12.3 we study the effect on a system containing clouds with a realistic mass spectrum, and study how star formation changes according to the relative position within the cocoon associated with the jet.

Most of the runs were evolved up to $\sim 10^7$ yr, while the jet was active and supplying energy to the cocoon. Only in one run (H4) was the jet switched off after

10^7 yr, and the further evolution of the system was followed until the cloud was completely destroyed.

12.2.4 Model of embedded cloud

We have chosen a model for the structure of the embedded clouds suggested by the numerical simulations performed by Baek *et al.* (2005), because the physical ingredients of these simulations are likely to be representative of the physical processes present in the ISM/IGM. The simulations have been devised to provide a reasonable model for clouds embedded within the IGM. Different sources of heating (UV-background, the wind and radiative flux from the central QSO itself) provide a significant energy input, which can promote the formation of pressure-confined clouds through thermal instability. Baek *et al.* (2005) have shown that, for typical IGM density and temperatures, similar to those considered in the present paper, the cooling time is a small fraction of the dynamical time, and the ISM is prone to the development of small, pressure-confined clouds. These are almost isothermal, with temperatures near the lower extreme of the cooling function ($T_{cl} \approx 10^4$ K). In the mass range $10^{2.5} \leq M_{cl} \leq 10^7 M_{\odot}$ they find a relationship between the truncation distance r_t and the total mass M_{cl}:

$$r_t = \lambda M_{cl,4}^{\beta}, \tag{12.3}$$

where we have defined $M_{cl,4} = M_{cl}/10^4 \ M_{\odot}$. If r_t is measured in pc, we obtain $\lambda = 28.87$, $\beta = 0.28 \pm 0.04$. The upper limit of this mass range corresponds to the Jeans mass for these clouds, implying that they are *not self-gravitating*.

As we have shown in Appendix A of Antonuccio-Delogu and Silk (2008), a reasonable model for these clouds is the *truncated isothermal sphere* (TIS), (Shapiro *et al.* 1999). Given the cloud's mass M_{cl}, the final parameters of the configuration will depend on the background ISM thermodynamic state, and we assume that the latter is described by an unperturbed ideal equation of state with density and temperature ρ_{cl} and T_{cl}, respectively, as appropriate to a high-temperature, low-density, fully ionized plasma (e.g. Priest 1987). We denote the solution of the TIS equation by $F(\zeta)$, so that the cloud density can be written as $\rho_{cl} = \rho_0 F(\zeta)$, ρ_0 being the normalization factor and $\zeta = r/r_0$ a normalized distance. The cloud's structure is entirely specified by assigning r_0, ρ_0 and the truncation distance r_t. Once the latter is determined by the radius–mass relation found by Shapiro *et al.* (1999), the two remaining factors are determined by imposing pressure equilibrium with the IGM, as shown in Appendix A of Antonuccio-Delogu and Silk (2008).

12.2.5 Star formation in the cloud

Although our maximum spatial resolution could allow us to resolve sub-parsec scales, we cannot follow star formation in any detail, because this would require the inclusion of more physics (for instance a very detailed treatment of molecular cooling) and temporal resolution a few orders of magnitude higher than the one we have adopted. Instead, we assume that the cloud is converting gas into stars at a rate determined by the Schmidt–Kennicutt law (Schmidt 1959, 1963; Kennicutt 1998) $\dot{\Sigma} = A\Sigma^n$, with $A = (2.5 \pm 0.17) \times 10^{-4} \, M_\odot \, \mathrm{yr}^{-1} \, \mathrm{kpc}^{-2}$, $n = 1.4 \pm 0.15$. At any time, we assume that star formation proceeds only within those regions of the simulation volume where the following criteria are satisfied:

(i) mass is larger than the Bonnor–Ebert mass;
(ii) temperature is less than a prescribed upper cutoff, i.e. we assume that star formation is sharply inhibited in regions having $T \geq T_c = 1.2 \times 10^4$ K.

The Bonnor–Ebert mass (Ebert 1955; Bonnor 1956) is defined as the largest mass that a pressure-confined, gravitating cloud can reach before becoming unstable:

$$M_{be} = 1.18 \frac{c_s^4}{\sqrt{G^3 p_{ext}}} = 23.55 \frac{T_4^2}{p_{ext}^{1/2}} \, M_\odot, \qquad (12.4)$$

where p_{ext} is the pressure at the surface of the cloud, temperature is expressed in units of 10^4 K, and we have assumed that an isothermal equation of state applies, so that the sound speed is given by $c_s = (k_B T/m_p)^{1/2}$. The temperature cutoff is often adopted in star formation models to exclude regions where the UV flux would be too high to permit significant star formation. Our criteria to select the star-forming region are very similar to those adopted by Schaye and Dalla Vecchia (2008).

From now on, all characteristic properties of the cloud, such as its mass or size, will always be referred to the *star-forming region* as just defined.

12.2.6 Issues about numerical resolution

As our main goal is to study in detail the evolution of star formation within the cloud, we need sufficiently high spatial resolution during each run. The outer parts of the cloud are stripped due to interaction with the jet, and it is in principle difficult to predict how much the cloud's size and mass will be reduced during the evolution. For this reason, we have applied a refinement criterion that will enable us to get high resolution even during the late phases, when the star-forming region is considerably reduced. The spatial resolution in FLASH is determined by three parameters: the number of blocks in the initial decomposition, which in 2D simulations is given by $N_{blx} \times N_{bly}$; the number of mesh cells within each

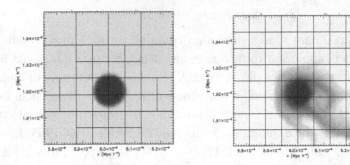

Figure 12.2 Computational block distribution for two different steps of run H1, taken at $t = 2.6 \times 10^5$ (*left*) and 5.3248×10^5 (*right*) (time in years). Each small square represents a FLASH block, and each of these contains 64 mesh cells (not reproduced in the figures). Note that the size of the region is $50h^{-1}$ pc, a small fraction of L_{box}, and the blocks represented are at the highest refinement level.

block, $n_{bx} \times n_{by}$; and the maximum allowed refinement level l_r. In the present work we have chosen: $N_{blx} = N_{bly} = 20$, $n_{bx} = n_{by} = 8$, and $l_r = 6$. Thus, the smallest spatially resolved scale (down to mesh level) along each direction is given by: $\Delta_x = \Delta_y = (L_{box}/N_{blx}) / (2^{(l_r-1)}n_{bx})$, which, for $L_{box} = 4h^{-1}$ kpc gives: $\Delta_x = 0.78125\,h^{-1}$pc. Note that the maximum refinement level around the cloud is reached at the start of the simulation, because of the refinement criterion adopted. The approximate number of mesh cells contained within a cloud of radius R_{cl}, n_m, can be estimated as $n_m = int\,(\pi R_{cl}2/\Delta A_m)$, where $\Delta A_m = \Delta_x \Delta_y$ is the area of a single mesh. In our case initially $R_{cl} \simeq 10h^{-1}$ pc, so we get $n_m \approx 514$. A typical block distribution is shown in Figure 12.2.

By default, FLASH refines the grid at those points where one of the components of the second spatial derivative of some user-selected quantities, normalized to the square of the spatial gradient, exceeds some pre-established threshold value. The quantities we check for refinement are density, pressure and temperature. This default criterion is sufficient to resolve the highly compressed regions on the scale of the clouds we are interested in, so we do not add any additional, customized refinement criterion.

12.2.7 Evolution of the cocoon

The injection of energy into the ISM/IGM can engender turbulence, mostly because the jet is supersonic at and near the injection point. Most previous work aimed at describing the general structure and evolution of the cocoon, however, has paid more attention to the global dynamics of the cocoon. Turbulence can have a significant impact on the evolution of the embedded clouds, and for this reason we study it in more detail in the next sections.

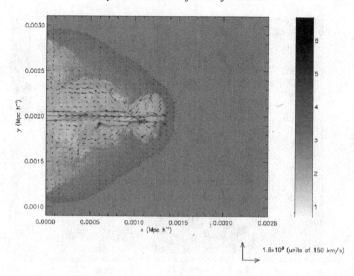

Figure 12.3 Density and velocity distribution at $t = 8.1 \times 10^4$ yr for run H4. The logarithmic density scale ranges from 10^{-3} to 10^3 e$^-$ cm^{-3}. The transition region between the cocoon and the external medium is threaded by a series of transonic shock waves (shocked ambient gas layer). The high-density enhancement within the cocoon originates out of the stripped material from a cloud initially set near $(0.6, 1.92)h^{-1}$ kpc, which has been shocked by the jet.

Soon after the jet enters the ISM/IGM a cavity forms, and the gas that has been swept out piles up into a transition layer. In Figure 12.3 we show an output of one of the runs at a rather advanced stage. We can easily distinguish an internal low-density, high-temperature cocoon and a *shocked ambient gas* region, externally bounded by a tangential discontinuity from the outer ISM/IGM. One of the most interesting features we find is that these regions are also *dynamically* very different. The region containing the shocked gas is threaded by a series of weak transonic shocks, and it has on average an expansion motion, while in the cocoon a large-scale circulation parallel and opposite to the main stream of the jet develops, originating from gas reflected away from the region near the hot spot. This circulation produces shear motions, which then decay into weak turbulence within the cocoon. Pre-existing clouds embedded within the ISM/IGM are heavily affected by this turbulence. The typical velocities of these shearing motions are large but, due to the very high temperatures within the cocoon (T $\approx 10^9$–10^{11} K), the motions themselves are only moderately transonic. As one can notice by inspecting Figure 12.4, the pressure within the cocoon can reach high values, because the temperatures are on average very high. Also, the region around the terminal part of the jet, near the hot spot, is subjected to high pressures, which increase steadily until the cocoon's expansion is halted by the ram pressure. We notice that the highest values of pressure are attained within the *shocked ambient gas region*, mostly driven by

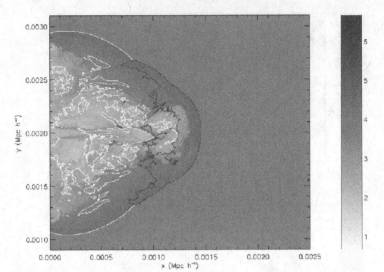

Figure 12.4 Density and pressure distribution for the same output as Figure 12.3. We show two pressure contours, corresponding to 5.77×10^9 (white) and 2.88×10^{10} (dark grey) K cm^{-3}.

the higher density, up to three orders of magnitude larger than the average density in the cocoon. The Mach number is higher near the jet, particularly near the injection point, but in the overall region (cocoon and shocked gas layer) it never reaches values larger than $\mathcal{M} \approx 3.0$–3.5. The dynamical evolution of the hot spot, i.e. the region between the tip of the jet and the terminal part of the cocoon, is quite interesting. Here, however, we concentrate on those features more directly related to the interaction with clouds, leaving to further work a more detailed analysis of the features of interest relevant to the modelling of the radio jet.

All runs were stopped when the jet was still active, except for run H4, which was continued for about 5×10^6 yr after the jet was switched off. This shutdown time was chosen to be $t \approx 10^7$ yr, within the range of a typical duty cycle for the AGN. The cocoon expands up to a maximum radius determined by the jet power and the environmental ram pressure. When the duty cycle of the AGN is completed, the jet injected power decreases very rapidly, and the cocoon is no longer fed by the jet. The evolution is mostly driven by inertial motions: a snapshot is shown in Figure 12.5. The typical velocities are still quite high within the cocoon, but the expansion of the shocked ambient gas region has already been slowed down by the ISM/IGM ram pressure, and it is now decelerating, while slowly expanding. Notice that the cloud is threaded by arc-like internal shocks within the cloud (Figure 12.7, *left*), previously noticed in similar simulations of shock–cloud interactions (Nakamura *et al.* 2006).

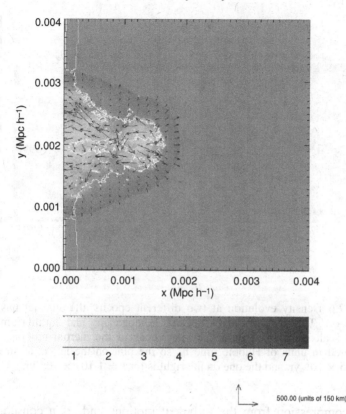

Figure 12.5 Density field of run H4 at $\Delta t = 2.45 \times 10^5$ yr after the jet has been shut off (i.e. at a time $t = 1.00245 \times 10^7$ yr since the beginning of the simulation). The cocoon and the tangential discontinuity are now dynamically very different, and their boundary is marked by a zero velocity contour.

The further evolution of the cloud during the passive phase shows some features not emphasized in previous papers. A continuous, decelerating flow now takes place within the cocoon, primarily driven by the inertial motions, because at the typical densities of the cocoon ($n_e \approx 10^{-4}$–$10^{-2}\,\mathrm{cm}^{-3}$) the gravitational pull is less than the inertial force. A zero total velocity line now separates the cocoon from the external ISM. In the latter the decelerating, outward-directed expansion motion changes to a wind as the plasma leaks out of the simulation box and the density decreases. A similar motion, but directed in the opposite direction, drives the gas past the left vertical boundary, i.e. towards the AGN and the galaxy bulge, increasing the density within this region. As a consequence of the generalized density decrease, the cooling time increases even further, although it was already much larger than the free-fall time-scale.

This backflow within the cocoon has some interesting consequences for the embedded clouds. During the active phase, the cloud was already subject to

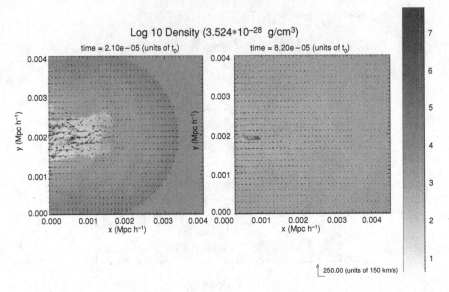

Figure 12.6 Density evolution at two different epochs after the jet has been switched off. The system evolves towards homogeneity and equilibrium on a temporal scale mostly determined by the decay of the inertial motions. Time is measured in units of Hubble time t_0, so the plot on the left is at an epoch $t = 2.835 \times 10^5$ yr, and the one on the right is for $t = 1.107 \times 10^6$ yr.

significant compression from the turbulent motions and, as a consequence, its density had also risen. Its temperature was, however, typically higher than $5 \times 10^5 - 10^6$ K, a level maintained mostly by the turbulent dissipation. However, during the passive phase, the cloud is embedded within a continuous decelerating stream, which is now more laminar than turbulent (Figure 12.6). This flow continues to compress the cloud, which cools efficiently: its average density also increases, and its temperature decreases, a trend that can be clearly seen on the right-hand side of Figure 12.7. On a longer timescale the cloud eventually is completely stripped, due to KH instabilities. Our simulations do not have the spatial and temporal resolution to further follow the fragmentation of this cloud, which will be treated in a separate paper. The overall action of the cloud's compression during the active jet phase, and of the cooling and further compression during the passive phase, can result into an occasional enhancement of the region of the cloud where favourable conditions for star formation are possible, as can be seen from Figure 12.7. The gas stripped away by KH instabilities from the cloud tends to form filamentary, high-density structures, where temperatures can reach high values ($T \sim 10^4 - 4 \times 10^5$ K). Star formation within these filaments, due to these high temperatures, is thus inhibited, and eventually the cloud is completely destroyed.

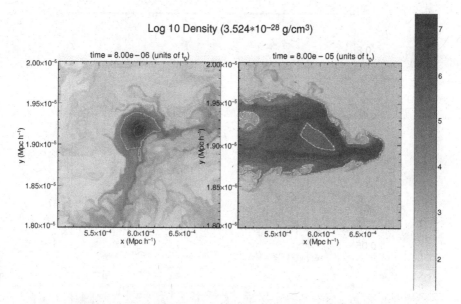

Figure 12.7 Evolution of density and temperature in the compressed cloud during the active phase (*left*), during which the jet is feeding energy into the cocoon, and some time after it has been switched off (*right*). The contours superimposed on the density maps trace regions of equal temperature, and correspond (from the inner- to the outermost contour) to the following temperatures: $T = 10^4$, 10^7 and 10^9 K.

12.2.8 Probability density distribution

The cocoon represents for the cloud an environment radically different from the hot ISM: on average, temperatures are higher and densities lower, and it is also in a turbulent state, permeated by a series of turbulent eddies, which produce a random, intermittent series of stresses on the cloud. This scenario is more complex than those envisaged in previous models of cloud evolution under external shocks (Klein *et al.* 1994; Nakamura *et al.* 2006), and a model of the turbulence within the cocoon is then a prerequisite to developing a model of the evolution of clouds within a turbulent environment, although this is beyond the purposes of the present work.

Simulations of turbulence in the ISM suggest that, on galactic scales, a log-normal probability distribution function (LNPDF) provides a viable description of the distribution of density fluctuations over a large range of densities typical of galactic ISM (e.g. Padoan *et al.* 1997; Wada and Norman 2001, 2007). The physical conditions of the system considered in this work, however, are very different from those typical of the galactic ISM, where temperatures are considerably lower ($T \approx 10\text{--}10^2$ K) and densities higher, so we do not expect *a priori* to find the same PDF. Moreover, the presence of the jet, which injects energy and momentum into

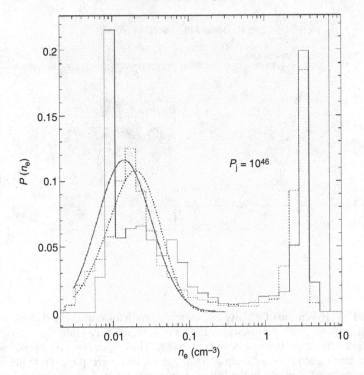

Figure 12.8 Density PDF for run H4. The analysis was performed only within the cocoon/shocked ambient gas region. The two histograms show the PDF at two different epochs: $t = 1.35 \times 10^4$ yr (dotted), and $t = 9.45 \times 10^4$ yr (continuous). The curves show two best fits for the latter epoch: an exponentially truncated power law (*continuous*) and a modified log-normal (*dotted*). The vertical line at $n_e = 7$ is the maximum density possible for fully radiative shocks (Bouquet *et al.* 2000). Best fit parameters are given by $(A, b, n) = (1.141 \times 10^5, 14.486, 24.612)$ (Equation 12.5), and $(B, \langle x \rangle, \sigma, b) = (3.66 \times 10^{-3}, -1.6, 0.498, 2.005)$ (see Equations 12.5 and 12.6).

an expanding cocoon, results in a *non-stationary* background. In Figure 12.8 we show the evolution of the density PDF for one of the runs (H4). One notices that the PDF is bimodal, with two peaks at densities lower and higher than the initial ambient density ($n_e = 1 \mathrm{cm}^{-3}$). These two peaks correspond to two spatially separated regions: the low-density distribution arises predominantly within the cocoon, while the high-density component is associated with the gas in the shocked gas region. The continuous and dotted curves are two least-square fits of the low-density PDF, with two different fitting functions: an exponentially truncated power law (continuous curve),

$$P_{\mathrm{tr}} = A|x|^n \exp\left(-\frac{x}{b}\right) \qquad (12.5)$$

and a modified log-normal (dotted curve):

$$P_{\text{lm}} = B \exp\left[-\frac{(x - \langle x \rangle)^2}{\sigma^2} - bx \right], \tag{12.6}$$

where $x = \log(\rho)$. The quality of the two fits is comparable: $\chi^2/dof = 1.76$ (truncated power law) and $\chi^2/dof = 1.32$ (modified log-normal). A modified log-normal distribution is generally expected in any fluid system whose density distribution is the result of a series of uncorrelated shocks (e.g. Vázquez-Semadeni 1994; Padoan *et al.* 1997; Passot and Vázquez-Semadeni 1998). Such a PDF has also been found in simulations of turbulence and global star formation in galactic discs (Wada and Norman 2001, 2007; Tasker and Bryan 2006), thus suggesting that it could also arise in stationary systems where turbulence is not only artificially forced within the simulation box. Note that in the model of Passot and Vázquez-Semadeni (1998) this particular form for the PDF should be independent of the dimensionality of the system, although the parameters characterizing the distribution will depend on it. It seems reasonable to assume that the main driver of this weak turbulence is multiple shocks within the cocoon, forming initially from the evolution of Kelvin–Helmholtz (KH) instabilities at the jet–ISM interface. These shocks then propagate within the cocoon, and are reflected at the internal interface with the shocked gas region.

12.2.9 *Evolution of star formation rate*

The influence of the jet/cocoon on star formation within the cloud is the result of the competition between different physical factors. The impact of the shock and the increased pressure within the cocoon result in an overall compression of the cloud, which tends to *increase* its density and specific SFR. On the other hand, the increase of temperature and stripping of the outer regions due to KH and other instabilities tend to reduce the mass of the cloud, thus contributing to a *decrease* of the global SFR.

The detailed temporal evolution of the cloud is governed by the turbulence within the cocoon, and subsequently by the evolution of the backflow. In Figure 12.9 we show the evolution of the *specific* and *global* SFRs during the active phase, when the jet is still feeding energy into the cocoon. One notices that in runs H0–H2 the specific SFR occasionally increases, due to the increase of the average density of the star-forming region. The *global* SFR also shows similar fluctuations, although of smaller amplitude, in all these runs. A similar trend is observed for the evolution of the mass of the star-forming region, Figure 12.10: some episodic increases are seen, in temporal and sign correspondence with the fluctuations of the SFRs. However, stripping tends to diminish the mass of the cloud, so we are forced

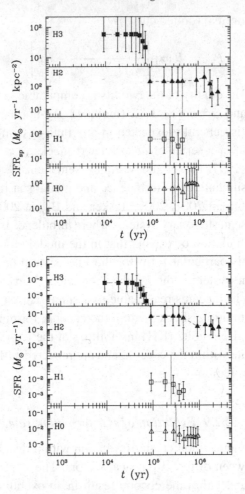

Figure 12.9 Evolution of the cloud's star formation. The red vertical bar labels the time when the jet/cocoon first hits the cloud. Error bars are computed from the $\pm 1\sigma$ errors of the Schmidt–Kennicutt law (Kennicutt 1998). *Upper plot*: specific star formation rate; *lower plot*: total star formation rate over the whole cloud.

to conclude that these episodic increases of the star-forming region can only be attributed to occasional cooling, driven by thermal instability. This gas is, however, soon warmed up again by the warm, turbulent environment of the cocoon, and the SFR then decreases.

A similar trend is observed in run H4, during the passive phase. We have already seen from Figure 12.7 that the cloud, exposed to the back-flow from the jet, becomes more filamentary and cold. However, we see from Figure 12.11 that its mass decreases dramatically, because the inner cold regions become less dense, and consequently both the specific and the global SFRs diminish. We can thus conclude that, in general, the interaction with the jet/cocoon tends to inhibit, in

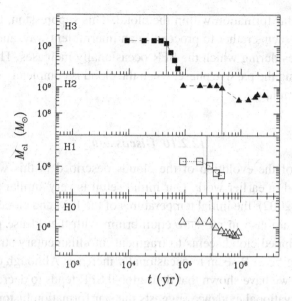

Figure 12.10 Evolution of the mass of the cloud's star-forming region.

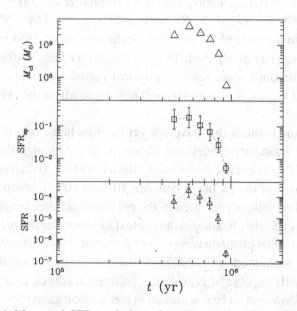

Figure 12.11 Mass and SFR evolution after the switch-off of the jet (run H4). Time is now measured in years after the switch-off epoch (arbitrarily fixed at $t = 10^7$ yr).

the long term, star formation within the cloud. This suppression, however, is not continuous, but seems rather to proceed in an intermittent way, and is interrupted by short episodes during which the SFR occasionally increases. This trend seems to continue during the passive phase, when the cloud is completely stripped and its SFR decreases rapidly.

12.2.10 Discussion

Some features of the evolution of the clouds described in this work have been previously found in earlier work. Our initial setup is very similar to that used by Mellema *et al.* (2002): the initial temperatures of the ISM and clouds are the same, and our clouds are also in pressure equilibrium with the diffuse phase. We also find that the stripped cloud seems to fragment into filamentary structures, whose density and specific SFR seem to occasionally increase. Although these structures seem to persist, we have shown that their global SFR tends to decrease.

Recent observational evidence suggests the star formation history in early-type galaxies can be more complex than in the standard, monolithic scheme. There is evidence both of quenching of star formation (Kaviraj *et al.* 2007; Schawinski *et al.* 2006) and of recent star formation episodes (Thomas *et al.* 2007; Yi *et al.* 2007; Kaviraj *et al.* 2008; Sarzi *et al.* 2008): both could be induced by AGN activity. It is then important to understand the physical mechanism of jet–ISM interaction, and how star formation can be affected. The simulations we have performed represent an attempt to introduce some realistic physical ingredients of the ISM structure into a numerical model, and to understand how these affect the evolution of SFR in the cloud.

There are some features that have not yet been included into this model. One is related to the quantitative relevance of this model. Although the duration and frequency of AGN jets are not well known (Blundell *et al.* 2002), typical estimates of the *active* phase of the jet range between 10^7 and 10^8 yr, comparable to those adopted in the simulations presented in the previous sections. Empirical evidence from modelling of SMBH feeding in the context of hierarchical structure formation suggests that the AGN phenomenon persists for several Gyr (Bromley *et al.* 2004; Haehnelt 2004; Mahmood *et al.* 2005), or many galaxy dynamical times. Hence AGNs are certainly capable of providing significant feedback over the timescales relevant to star formation and gas accretion in hierarchical galaxy formation models.

Our simulations are restricted to only *one episode* of AGN activity: a further episode of jet injection would certainly propagate into a very hot and low-density ISM, because our simulations show that the diffuse ISM/IGM within the cocoon reaches very high temperatures ($T \approx 10^8 - 2 \times 10^{11}$ K) and low densities ($n_e \approx 10^{-2}$–10^{-1} cm^{-3}), depending on the initial temperature and on the jet power at the time of its maximum expansion. Under these conditions, the cooling time exceeds the dynamical time by a factor $t_{\text{cool}}/t_{\text{dyn}} \approx 6.5 \times 10^2 - 3 \times 10^5$, thus the heated gas does not cool significantly (as already noted by Inoue and Sasaki (2001)). Thus, a second jet would propagate into an environment where the ram pressure will be considerably higher, $p_{\text{ism}} \approx 10^6 - 2 \times 10^{10} \, \text{K cm}^{-3}$ (the initial pressure in the ISM was $p_{\text{ism}} = 10^7 \, k^{-1}$), and would probably be quenched. We shall verify these predictions in subsequent work.

Our simulations differ significantly from those of Kaviraj *et al.* (2007), because the turbulence within the cocoon surrounding the propagating jet is generated naturally by the interaction of the jet with the ISM/IGM, while Kaviraj *et al.* studied the turbulence associated with the disruption of pre-existing ISM/IGM clouds due to KH instabilities at the jet–IGM interface. There are, however, some features that we have not addressed in our simulations. Sutherland and Bicknell (2007) have simulated the propagation of a jet that first emerges out of a gaseous disc (see also the contribution of Bicknell *et al.* in this volume, Chapter 14). They note that the disc has a strong effect on the energy budget of the jet and also on its morphology: during its initial propagation the jet percolates through the disc, and it needs more time to overcome this initial pressure and emerge out of the disc into the ISM. Sutherland and Bicknell do not consider the compression of pre-existing clouds: the main target of their work is a detailed study of the morphology of the emerging jet, and a comparison with observations. A significant gaseous component, often distributed in disc-like structures, seems, however, to be present in relevant amounts ($10^9–10^{10} \, M_\odot$) also in a fraction of early-type galaxies (Goudfrooij *et al.* 1994; Macchetto *et al.* 1996; Morganti *et al.* 2006; Combes *et al.* 2007).

12.3 Global quenching in elliptical galaxies

12.3.1 Introduction

Active galactic nuclei (AGN) have been advocated in recent years as sources able to influence the evolutionary history of stellar populations within their host galaxies. Here we present a physically motivated model for the quenching of star formation in high redshift ellipticals, starting from a series of numerical simulations.

12.3.2 Multicloud simulation

As in Section 12.2, to perform the simulation, we used FLASH v.2.5 (Fryxell *et al.* 2000), a parallel adaptive mesh refinement (AMR) code, which implements a second order, shock-capturing PPM solver. Again we include radiative cooling, using the standard cooling function from Sutherland and Dopita (1993), conveniently extended towards higher temperatures, $T > 10^7$ K, which are attained within the cocoon generated in this simulation (see Appendix A in Antonuccio-Delogu and

Silk 2008). We also take into account gravity, but we neglect thermal conduction and large-scale, ordered magnetic fields.[1]

The jet is modelled as a one-component fluid, characterised by a density ρ_j, which is a fraction ϵ_j of the initial density of the interstellar medium. In order to prevent numerical instabilities at the jet/ISM injection interface, we use a steep, continuous and differentiable transverse velocity and density profile, as we did before (Antonuccio-Delogu and Silk 2008; Perucho et al. 2004, 2005):

$$v_{x,j} = \frac{V_j}{\cosh\left\{\left(\frac{y-y_j}{d_j}\right)^{\alpha_j}\right\}}, \tag{12.7}$$

$$n_j = n_{env} - \frac{(n_{env} - n_j)}{\cosh\left\{\left(\frac{y-y_j}{d_j}\right)^{\alpha_j}\right\}}, \tag{12.8}$$

where $\alpha_j = 10$ is an exponent that determines the steepness of the injection profile, n_j, n_{env} denote the jet and environment electron number densities and the scale length d_j characterizes the width of the jet. The power injected by the jet is then given by $P_j = \beta 2\pi d_j^2 \rho_j V_j^3$, with $\beta \simeq 0.7158$. In this simulation we set $P_j = 10^{46}\,\mathrm{erg\,s^{-1}}$.

We model the environment where the jet propagates as a two-phase interstellar medium comprising a hot, diffuse, low-density component and a cold, clumped system of high-density clouds in pressure equilibrium with the diffuse component. The warm phase is characterized by a density profile $\rho_{env}(r)$ and a constant temperature T_{env}. We assume that the diffuse gas is embedded within a dark matter (DM) halo, the latter being described by a NFW density profile. This DM halo is chosen to have a total mass $M_h = 5 \times 10^{11}\,M_\odot$, concentration $c = 10.2$ and a scale length $l_h = 206\,h^{-1}\,\mathrm{kpc}$. The ISM gas is assumed to be a fraction $m_g = \Omega_g/\Omega_{DM} \approx 0.212$ of the DM, and to be distributed in hydrostatic equilibrium within this DM halo, following the prescription given in Appendix C of Hester (2006). Each of the clouds in the cold component is modelled as a truncated isothermal sphere (TIS: Shapiro et al. 1999; Iliev and Shapiro 2001), because this model seems to adequately reproduce the properties of clouds formed in simulations of a thermally unstable ISM. TIS spheres possess a finite radius, and are characterized by two parameters, which we assume to be the mass M_{cl} and a typical radius r_{cl}. They are exact solutions of the equilibrium equations for isothermal spheres confined by an external pressure. We distribute 300 clouds within the simulation volume, using a mass spectrum previously derived from numerical simulations (Baek et al. 2005). Summarizing, the simulation is specified by ten parameters: three of these describe

---

[1] Even an initially weak, small-scale, tangled magnetic field would be amplified to a level that inhibits thermal conduction, by two to three orders of magnitude with respect to the classical Spitzer values.

Table 12.3 *Parameters of the simulation*

Parameter	Adopted value
$M_{\rm h}$	5×10^{11}
c	10.2
$l_{\rm h}$	206
$m_{\rm g}$	1.06×10^{11}
$n_{\rm env}$	10^2
$T_{\rm env}$	10^7
$n_{\rm j}$	1
$d_{\rm j}$	10
$P_{\rm j}$	10^{46}

Mass is measured in units of $M_\odot$, distances are measured in h^{-1} kpc, $n_{\rm env}$, n_j are the central gas electron and jet densities, respectively, in units of cm^{-3}, $T_{\rm env}$ is in K, d_j and P_j are the jet width and power, respectively. The latter is expressed in cgs units.

the DM halo ($M_{\rm h}$, c, $l_{\rm h}$), two the diffuse phase of the ISM ($m_{\rm g}$, $T_{\rm env}$), two the mass distribution within the cold component of the ISM ($M_{\rm cl}$, $r_{\rm cl}$), and the last three describe the jet ($n_{\rm j}$, $d_{\rm j}$, $P_{\rm j}$).

We choose a simulation box having a size $L_{\rm box} = 40h^{-1}$ kpc, so that the jet will diffuse through it at the end of the simulation. The spatial resolution attained is a function of the maximum refinement level and of the structure of the code. For a block-structured AMR code like FLASH, where each block is composed of $n_x \times n_y$ cells, the maximum resolution along each direction is given by $L_{\rm box}/(n_x 2^l)$, where l is the maximum refinement level. In this simulation: $n_x = n_y = 8$ and $l = 6$, thus the minimum resolved scale is 78.125 pc. Note that we are performing a 2D simulation, but we do not impose any special symmetry.

We assume that the duty cycle of the jet is active for a time of 4×10^6 yr, as is typical for jets having such power (Shabala *et al.* 2008), and that its power declines linearly after this epoch until the jet switches off at $t \approx 2 \times 10^7$ yr. As for the SFR, we assume that the clouds are converting gas into stars with a rate specified by the Schmidt–Kennicutt (SK) law (Schmidt 1959, 1963; Kennicutt 1998): SFR $= \dot{\Sigma} = A\Sigma^n$, where $A = (2.5 \pm 0.17) \times 10^{-4}\ M_\odot\ {\rm yr}^{-1}\ {\rm kpc}^{-2}$ and $n = 1.4 \pm 0.15$. We further require that SF is ongoing only within regions having a mass larger than the Bonnor–Ebert mass (Ebert 1955; Bonnor 1956), and temperature less than a specified upper limit, i.e. $T \leq 1.2 \times 10^4$ K. These are rather conservative constraints, which tend to underestimate the magnitude of the negative feedback. Our initial setup is very different from that of the very recent simulation by Sutherland and Bicknell (2007), particularly in one very important point: while we model a population of pressure-confined clouds *embedded* within the ISM, Sutherland and Bicknell (2007) put a turbulent disc around the jet's source, embedding AGN in it, as they are more interested in addressing problems related to the evolution of GPS and CSS radio sources. Also, note that their simulation box is much smaller than ours (1 kpc vs. 54 kpc), and the temporal scale is also very different ($\sim 10^5$ yr in their simulation, compared to 2×10^7 yr in the present work). Moreover, one may notice that we have neglected the velocity field of the inhomogeneous component: in fact, as already noted by Sutherland and Bicknell (2007), for the spatial and temporal scales of interest, its inclusion has little effect on the evolution of the clouds and of their SF. The typical turbulent velocities of clouds are in the range 10–70 km s^{-1}. Thus, on typical timescales $\sim 2 \times 10^7$ yr in our simulation, these clouds have displacements $l \sim 0.2$–1.4 kpc, smaller than the *average* distance between clouds.

12.3.3 *Global quenching of star formation*

The global evolution show that one can distinguish two main phases corresponding to the evolution of the cocoon: an *active* and a *passive* phase. While the jet is active, the cocoon is fed and it expands almost self-similarly, while when it is switched off, the cocoon diffuses and affects a larger fraction of the ISM.

Initially, the interaction mostly takes place through the weak shock at the interface between the cocoon and the ISM. This shock compresses and heats up the clouds, having two opposite effects on SF: compression tends to increase SF, but the larger temperature also increases the critical mass for gravitational collapse, thus decreasing the volume fraction of the clouds that can actively form stars. Globally, we observe an initial enhancement of SFR (i.e., we observe a *positive feedback*). The effects of the two concomitant processes on clouds can be seen in Figure 12.12, where we build the shell-SFs by adding the SF of clouds lying in a concentric annulus of mean radius r and plot them as a function of radius for different time steps. Here, few episodes of positive feedback are evident in the clouds that are nearer to the cocoon. Later, the cocoon propagates within the medium inducing a general increase of the temperature, which heats up the clouds and decreases their density, and finally the outer regions of the clouds are stripped due to KH instabilities, as we studied in more detail in Antonuccio and Silk (2008). These effects tend to reduce the mass of the clouds, decreasing SF, and eventually drastically suppressing SF on a timescale of 2–3×10^6 yr (*negative feedback*). Obviously, the more external regions in the galaxy are affected later by the disruptive effect of the cocoon, being almost unperturbed until $t \sim 10^6$ yr and $t = 1.6 \times 10^6$ yr, respectively at $r \sim 30$ and ~ 45 kpc (see Figure 12.12). As we see from this simulation, approximately on a timescale over which the cocoon diffuses, the clouds are destroyed: this is

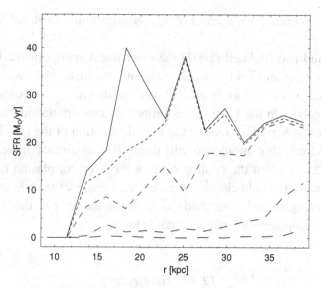

Figure 12.12 SFR in circular (equal-spaced) shells as a function of radius of shell. The results for different time intervals are shown, from continuous to long dashed lines spanning $t = 0.4, 1, 1.6, 2.5, 3.2$ Myr.

then the typical timescale over which SF will be inhibited. Obviously, one must observe that this timescale is not exactly coincident with the duty cycle of the jet.

Following Antonuccio and Silk (2008), we restricted our simulation to a single episode of jet injection. Any subsequent event of jet injection would have little influence on SF, since the jet would propagate through a high-temperature ($T \sim 10^8$–10^{11} K) and low-density ($n_e \sim 10^{-2}$–10^{-1} cm^{-3}) environment. As already noted by Inoue and Sasaki (2001) and in Antonuccio and Silk (2008), after the injection of the jet, the cooling time of the diffuse ISM exceeds the dynamical time by a factor $6.5 \times 10^2 - 3 \times 10^5$; thus the heated gas is not able to cool, quenching the second emitted jet. The extent of this region crucially depends on the physical parameters of the medium and on P_j. The case we present in this section is that of a very powerful jet, so at the end all the gas within the affected region is influenced by the expansion of the cocoon. For less powerful jets, it is reasonable to expect that the extent of the quenched star-forming region will be much smaller.

We should, however, remark that this scenario may not be exhaustive of all the possibilities. In a recent paper, Krause and Alexander (2007) studied the evolution of clouds lying very near to a jet. The clouds are destroyed by the KH instability produced by the jet, but the authors notice that a fraction of these clouds, under the effect of thermal instabilities, evolve into filamentary structures that are characterized by high densities. These filaments could then still host SF, although it is not clear what the *quantitative* relevance of this phenomenon is to global SF within the host galaxy.

In Sutherland and Bicknell (2007) the mechanical energy of the jet is mostly dissipated when the jet tries to percolate through the disk, while when the jet has already formed a cocoon, as in our simulation, its energy feeds the turbulence within the cocoon. On the large scales probed by our simulation, the turbulence within the cocoon is responsible for the final destruction of the clouds. When the jet is switched off, this turbulence will decay but, as already noticed by Inoue and Sasaki (2001), often the cooling time of the very hot plasma is too long to allow the formation of cold clouds. Other external episodes such as small mergers could, however, again fuel some cold gas, and strip the hot gas, thus favouring the subsequent development of embedded cold clouds.

12.3.4 Timescales

Although we discuss the output of only one single simulation, the general model of jet propagation in the ISM that is also probed by our simulation allows us to determine one of the most relevant parameters of the star feedback model, which we will develop in the next paragraphs: the typical timescale for suppression of stellar formation.

In the self-similar expansion model of Kaiser and Alexander (1997), the cocoon is supposed to propagate into an unperturbed ISM, which is well described by a power-law profile, $\rho(r) = \rho_0(r/a_0)^{-\beta}$, and the typical scale length of the jet varies with time according to

$$L_{\mathrm{j}} = c_1 a_0 \left(\frac{t}{t_0}\right)^{3/5-\beta} , \tag{12.9}$$

where (Kaiser and Alexander 1997, Equation 5)

$$t_0 = 1.186 \times 10^6 \left(\frac{a_0^5 \, \rho_0}{P_{\mathrm{j},45}}\right)^{1/3} \mathrm{yr} \tag{12.10}$$

and $P_{\mathrm{j},45}$ is the jet's mechanical power in units of $10^{45} \, \mathrm{erg \, s^{-1}}$.

We have assumed that the number density of cold, star-forming clouds is proportional to that of the diffuse gas, and thus can be well approximated by the same power-law density profile outside the core ($r > a_0$). There exists a threshold cloud number density under which SF becomes negligible, and we suppose that this corresponds to a value of the gas density $\rho_{\mathrm{cr}} \approx 10^{-27} \mathrm{g \, cm^{-3}}$, which is reached at a distance $r_{\mathrm{cr}} = a_0(\rho_0/\rho_{\mathrm{cr}})^{1/\beta}$. Inserting this into Equation 12.9, we find that the

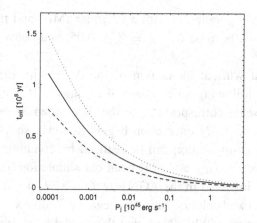

Figure 12.13 Timescale for feedback. The three curves refer to the same density profile, having $\rho_0 = 1.7 \times 10^{-25}\,\mathrm{g\,cm^{-3}}$, $\beta = 2$, and three values for $a_0 = 0.8,\ 1,\ 1.2\ h^{-1}$ kpc (dashed, continuous and dotted curves, respectively). The strong dependence on a_0 is simply a consequence of the high exponent with which it appears in the expression for t_0.

characteristic timescale needed to reach this distance is given by

$$t_{\mathrm{fb}} = t_0 \left(\frac{\rho_0}{\rho_{\mathrm{cr}}} \right)^{\frac{5}{\beta(3-\beta)}} c_1^{-\frac{5}{3-5\beta}}. \tag{12.11}$$

We plot in Figure 12.13 the dependence of t_{fb} on jet power P_{j}. As we can see, typical values for the timescale of feedback are in the range 10^6–10^8 yr, for reasonable values of the input parameters. The values of P_{j} that we choose are representative of the observed mass range of black holes at the centres of typical galaxies, according to the empirical relation between M_{bh} and P_{j} found by Liu *et al.* (2006). Thus, the quenching timescale for our simulation, of the order of a few Myr, will increase by about two orders of magnitude for less powerful jets.

12.3.5 Synthetic colours

The simulation discussed in this paper describes the impact that the jet emitted by AGN has on quenching SF. However, we would like to transpose this modification of the SF into a change of some observable quantities. Colours and absorption/emission lines are the most important observables: here we concentrate on the former. In order to reproduce galaxy colours, we use the synthetic models of Bruzual and Charlot (2003), which allows us to encompass a wide range of metallicities, starting from galaxy ages t_{gal} of 10^5 yr, and gives a full coverage in wavelength from 91 Å to 160 μm. We start from a single initial burst model, consisting of a single population with a Chabrier IMF[2] and three values of the metallicities: $Z = 0.008$, 0.02 (i.e. $Z = Z_\odot$), 0.05, and convolve it with a SFR law.

We suppose that without interaction of the AGN with the ISM, SF within the clouds evolves following an exponential SF law, SFR $\propto e^{-t/\tau}$, where τ is a characteristic timescale corresponding to the time when SF is reduced by e^{-1}. Then at $t = t_{AGN}$, the AGN interaction begins to work, shocking the ISM and inhibiting SF. This configuration can be realized by combining the unperturbed exponential SFR for $t \leq t_{AGN}$ and SFR from our simulation for $t > t_{AGN}$. Due to the short timescale involved in the AGN effect on SFR, positive feedback is not observable in the derived colours, while the negative feedback is directly translated into a strong truncation of SFR. Note that this model is an approximation, since less powerful AGNs are observed at low redshifts and the decrease of SF can be shallower, as discussed in the previous section (see, also, the parametric model of SF quenching discussed in Martin *et al.* (2007)). If $\tau \to \infty$, the SFR becomes a burst with finite length t_{AGN} (i.e. constant up to t_{AGN} with a null value for $t > t_{AGN}$).

We calculate colours by convolving the filter responses of u, g, r, i and z SDSS and the FUV and NUV GALEX bands (Martin *et al.* 2005; Yi *et al.* 2005; Kaviraj *et al.* 2007) with synthetic spectra. In Figures 12.14, 12.15 and 12.16 we show the change with time of colours $FUV - NUV$, $NUV - g$ and $g - r$ and the effect of quenching. In particular, in Figure 12.14 we compare the colours of our inhibited SF (with $t_{AGN} = 1$ and $5\,\mathrm{Gyr}$) and two simple models using an exponential SF with $\tau = 0$ and $\tau = 1\,\mathrm{Gyr}$. SF inhibition essentially reddens the colours starting at $t > t_{AGN}$. Ultraviolet is very sensitive to SF inhibition (see the middle panel in Figure 12.14), with variations of 1–3 magnitudes in $NUV - g$; less sensitive are the visible colours, with variations of ~ 0.2–0.4 in $g - r$. The colour $FUV - NUV$ increases after the action of the AGN up to ~ 1–4, depending on t_{AGN}, but after this event it decreases, reaching a value ~ 1 for large t_{gal}. Therefore, the latter colour is degenerate, because one is not able to univocally determine the galaxy age from colour (each colour corresponds to two possible values of t_{gal}).

A single burst model ($\tau = 0$) is similar to our models with inhibited SF for $t > 2$–$4\,\mathrm{Gyr}$, but predicts larger values for $t < 2$–$4\,\mathrm{Gyr}$. While $g - r$ colours for the different models shown in Figure 12.14 are similar after $t \sim 6$–$8\,\mathrm{Gyr}$, the other colours are comparable at larger ages $t \sim 8$–10 Gyr, being more sensitive to SF. Not shown in the figure is the case $\tau \to \infty$, characterized for fixed values of t_{AGN} and t_{gal} by bluer colours.

---

[2] Galaxy colours are unchanged if we use a Salpeter IMF, since these two IMFs differently describe the distribution of low-mass stars that contribute little to the light distribution, while strongly affecting the total stellar mass (Tortora *et al.* 2009).

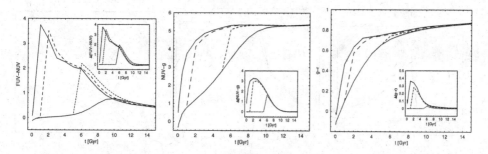

Figure 12.14 Estimated synthetic colours $FUV - NUV$, $NUV - g$ and $g - r$ (in AB system) as a function of galaxy age (in units of Gyr) for $Z = Z_\odot$. Red, blue, green and orange curves correspond respectively to an unperturbed exponential SF with $\tau = 0$, an exponential SF with $\tau = 1\,$Gyr and $t_{AGN} = 1\,$Gyr, an exponential SF with $\tau = 1\,$Gyr and $t_{AGN} = 5\,$Gyr and an unperturbed exponential SF with $\tau = 1\,$Gyr. In the inset panels we show the difference in colours with respect to the reference one with an unperturbed exponential SF with $\tau = 1\,$Gyr as a function of galaxy age; the colour code is the same as in the plot of colours vs. age (Figure 12.15). For colour version see http//:www.cambridge.org/9780521192545.

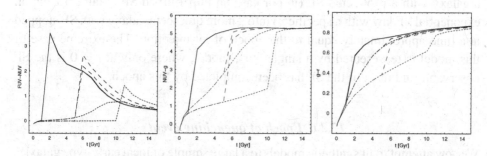

Figure 12.15 Estimated synthetic colours $FUV - NUV$, $NUV - g$ and $g - r$ (in AB system) as a function of galaxy age (in units of Gyr) for $Z = Z_\odot$ and for $\tau = 1\,$Gyr (red lines) and $\tau = 3\,$Gyr (blue lines). Continuous, long-dashed and short-dashed lines correspond to $t_{AGN} = 1, 5, 10\,$Gyr. For colour version see http//:www.cambridge.org/9780521192545.

These considerations are obviously dependent on the choice of τ and Z. In Figure 12.15, we compare the results for $\tau = 1$ and $3\,$Gyr. At fixed age, a more protracted unperturbed SFR predicts smaller $NUV - g$ and $g - r$ and a larger $FUV - NUV$. Metallicity has the opposite effect: $FUV - NUV$ decreases if we increase metallicity, while $g - r$ increases; finally, a more complex behaviour

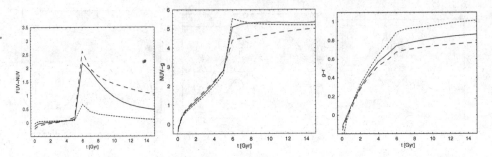

Figure 12.16 Estimated synthetic colours $FUV - NUV$, $NUV - g$ and $g - r$ (in AB system) as a function of galaxy age (in units of Gyr) for $\tau = 1$ Gyr and $t_{AGN} = 5$ Gyr. Long-dashed, continuous and short-dashed lines correspond respectively to $Z = 0.008$, 0.02, 0.05. For colour version see http//:www.cambridge.org/9780521192545.

is observed for $NUV - g$, as can be seen by inspecting Figure 12.16. These complex dependencies on metallicity are linked to the details of stellar population prescriptions.

We can outline a picture where the AGN effect has a main role in the evolution of brighter elliptical galaxies. In particular, inhibition of SF by AGN transforms a galaxy with a protracted SF (in our case an unperturbed SF obtained using an exponential SF law with a specific τ) into a more quiescent galaxy with SF stopped at a time approximately equal to the epoch of jet injection. The extreme case of this model is represented by a single burst model, where only at $t = 0$ is the SF observable and most of the SF has been completed by this epoch.

12.3.6 Epoch of quenching event

We now attempt to fit synthetic models to a large sample of local early-type galaxies (ETGs), in order to obtain information about the epoch of quenching event and recent star formation (RSF). The galaxy sample is presented in Section 12.3.7, while the spectral fit procedure and the first results are described in Section 12.3.8. We go into more detail in Sections 12.3.9 and 12.3.10, where quenching and the properties of the recently formed stellar populations are discussed. Finally, a few comments on galaxy evolution are addressed in Section 12.3.11.

12.3.7 Galaxy sample

We use a sample of ETGs extracted from the SDSS, using a selection procedure described in Kaviraj *et al.* (2007). The initial selection is made using the *fracDev* parameter in SDSS, which attributes a weight to the best composite (deVaucouleur's + exponential) fit to the galaxy image in a particular band. The criterion *fracDev*

> 0.95 has been proven to be extremely robust, allowing one to pick up ~90% of ETGs in a typical sample of SDSS galaxies. Furthermore, a visual inspection of SDSS images is needed to refine the selection: the ability to classify galaxies obviously depends on redshift and apparent magnitude of the observed galaxies. In order to construct a magnitude-limited sample, we restrict ourselves to r-band magnitude < 16.8 and redshift z < 0.08. Finally, cross-matching with ultraviolet GALEX data produces the final sample, with measured magnitudes in SDSS bands u, g, r, i, z and GALEX FUV and NUV (see Kaviraj *et al.* (2007) for further details). In this way, we have a wide coverage of galaxy spectra, since in the optimal cases, magnitudes in seven bands are observed, ranging from $\lambda \sim 1500\,\text{Å}$ up to $\lambda \sim 9000\,\text{Å}$. SDSS magnitudes have typical uncertainties of ~0.01, while GALEX data are more uncertain with mean errors of ~0.25 and ~0.15, respectively for FUV and NUV magnitudes.

In Figure 12.17 we show a colour–colour ($NUV - r$ vs. $g - r$) diagram from our synthetic tracks. In the inner plot we present our sample, superimposing the galaxies, colour coded according to redshift bins. This diagram gives information about both t_{gal} and t_{AGN}, selecting specific values for two colours and the best parameters of galaxies. Different values of galaxy parameters predict galaxy colours that populate different regions of the diagram.

The sample of Kaviraj *et al.* (2007) is suitable for our analysis, since AGN spectral features are observed in the spectra of many galaxies. Type-I AGN are automatically removed by using the SDSS spectral classification algorithm, while we are interested in galaxies that host a Type-II AGN. To distinguish between normal star-forming galaxies and galaxies hosting an AGN, it is usual to analyse a few strong emission lines, e.g. the emission line ratios [OIII/Hβ] and [NII/Hα] (Baldwin *et al.* 1981; Kauffmann *et al.* 2003). Galaxies with both these indicators measured represent ~65% of the galaxy sample, and ~86% of them have spectral features consistent with those of LINER, Seyfert or transition objects.

Emission from Type-II AGNs does not affect the stellar continuum of host galaxies. In fact, for luminous Type-II AGNs the maximum contamination in flux amounts to a small percentage in visible bands and to less than 15% (translating to 0.15 mag) in UV bands (Kauffmann *et al.* 2003; Salim *et al.* 2007). In addition, galaxies hosting AGNs are systematically redder in the UV colours than their counterparts that do not have AGNs, further suggesting that continuum emission from AGNs leaves galaxy spectra unaffected.

12.3.8 Spectral fit

We build a library of synthetic spectra, using our SFR prescription with t_{gal}, τ, t_{AGN} and Z as free parameters. We divide the sample into three redshift bins 0–0.04, 0.04–0.06 and 0.06–0.08 and we move our spectra to the median redshift

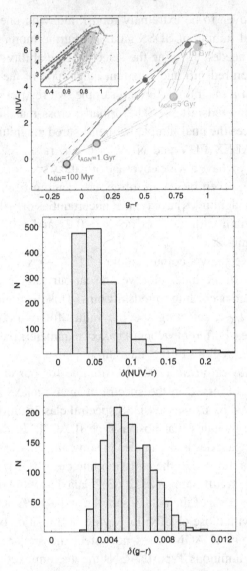

Figure 12.17 UV/optical colour–colour diagram ($NUV - r$ vs. $g - r$) from synthetic spectra redshifted to median redshift of the sample $z_{med} = 0.057$ assuming $\tau = 1$ Gyr and metallicities $Z = 0.008$ (long-dashed line), $Z = 0.02$ (continuous line) and $Z = 0.05$ (short-dashed line). Red, blue, green and yellow lines correspond, respectively, to $t_{AGN} = 0.1$, 1, 5, 10 Gyr. The large points set the values of t_{AGN} on each synthetic track, while the small ones indicate t_{gal} ($= 0.1$, 1, 5, 10 Gyr), which on different tracks correspond to different sets of colours after the quenching. In the inset panel we superimpose the sample galaxies on theoretical tracks, with violet, grey and cyan points indicating galaxies in the redshift bins 0–0.04, 0.04–0.06 and 0.06–0.08. Instead of error bars, we show in the lower panels the distributions of uncertainties on galaxy colours $g - r$ and $NUV - r$. For colour version see http//:www.cambridge.org/9780521192545.

of each bin (respectively 0.031, 0.052, 0.069). Synthetic magnitudes and colours are obtained by convolving these redshifted spectra with the filter responses. Finally, the synthetic colours are fitted to the observed ones ($FUV - NUV$, $NUV - g$, $u - g$, $g - r$, $g - i$ and $g - z$), by a maximum likelihood method, which allows us to estimate the best values for the free galaxy parameters. In this way, we do not need to have K-corrections, leaving the fit safe from possible uncertainties introduced by these corrections, and the simplistic division of our sample into only three redshift bins will not affect our estimates. Later, we will discuss the K-correction that we derive from the fitting procedure and will use to obtain rest-frame magnitudes.

We restrict the synthetic library to spectra with t_{gal} free to change (in tiny steps) up to 17 Gyr, $t_{AGN} \in (0, 15)$ Gyr, $Z \in (0.008, 0.02, 0.05)$ and three reference values for the SFR scale $\tau = 1, 3, 5$ Gyr. For each value of t_{gal} we have more than 150 spectral models. This library is wide enough to reproduce spectral features of ETGs (Panter *et al.* 2008). The range of metallicities used here has been shown to be representative of luminous ETGs with $M_B \simeq 19$ (Romeo *et al.* 2008; Tortora *et al.* 2009). In particular, the fit of spectra with an unperturbed exponential SFR and variable Z to local ETGs gives on average values of $\tau \sim 1$ and $Z \geq Z_\odot$ (Tortora *et al.* 2009). Such an unperturbed exponential SF with a low timescale τ can disguise a more complex SF evolution, which we have obtained from the simulation and want to probe against observations.

Before proceeding to model the properties of the observed sample of galaxies, we briefly discuss the possible systematics that the fitting procedure can generate in the best-fit parameters, by performing a set of Monte Carlo simulations on synthetic colours. We extracted a large sample of simulated spectra from our SED library with random t_{gal}, t_{AGN}, τ and Z. Then, we applied our fitting procedure and compared the recovered best fit parameters against the intrinsic ones. While t_{gal}, t_{AGN} and Z are recovered quite well,[3] τ is poorly constrained. To reduce the unavoidable degeneracies in the fitting procedure, it would be acceptable to set τ to a fixed value, however, in the following we will discuss the cases with τ both variable and constant. In addition, to further reduce the *noise* in our final results we will also analyse the effect of adopting constant metallicity.

Coming back to the observed sample of galaxies, we use different synthetic libraries to fit the observed colours, extracting spectra from our initial and extended set of synthetic models. We show the results of this analysis in Figure 12.18, where we plot the distribution of recovered values of $t_{gal} - t_{AGN}$. If on the one hand the change of spectral library can modify the estimates of single values of the parameters, on the other hand it leaves the main clump of the distribution of

---

[3] Note that the input value for t_{AGN} is correctly recovered if $t_{gal} - t_{AGN} \geq 0$, while for $t_{gal} - t_{AGN} < 0$ it is not possible to constrain its value.

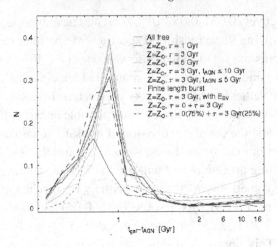

Figure 12.18 Distribution of recovered best fit values of $t_{gal} - t_{AGN}$ for different spectra sub-libraries extracted from our initial library. On the y-axis, we plot the fraction of galaxies (normalized to dimension of the sample) within bins of $t_{gal} - t_{AGN}$. In the range $t_{gal} - t_{AGN} \in [0-2]$ Gyr the scale is linear, while for $t_{gal} - t_{AGN} > 2$ Gyr we use an arbitrary logarithmic scale. In the linear range we group galaxies in bins of size 0.2 Gyr, while larger bins are used outside this range. In particular, we show the results leaving all parameters free to change, setting $Z = Z_\odot$ and $\tau = 1, 3$ or 5 Gyr; for $\tau = 3$ Gyr we analyse two cases with constraints on t_{AGN} and then include internal extinction in the fit. We analyse the case of a finite length burst ($\tau = \infty$) and a combination of two coeval stellar populations with $\tau = 0$ and $\tau = 3$ Gyr with a free mass ratio or a fixed one. See labels in the plot and discussion in the text for further details.

differences $t_{gal} - t_{AGN}$ unaffected. Due to the correlations among parameters and constraints imposed on some of them, the sample distributions of recovered values for t_{gal}, t_{AGN} and Z can change, but median values for $t_{gal} - t_{AGN}$ (~ 0.8–0.9 Gyr with a median scatter of ~ 0.1–0.2 Gyr) are almost unchanged, with a scatter among the different estimates consistent with 0. If we impose strong constraints on t_{AGN} (e.g. $t_{AGN} \lesssim 5$ Gyr), we observe a departure from the mean trend obtained using other libraries, with a peak at very large values of $t_{gal} - t_{AGN}$ (corresponding to quenching at high redshift), but this constraint does not seem to be motivated since it decreases the quality of fit. In addition, note that for $\tau = 1$ Gyr the distribution peak is slightly lower and less prominent, mainly due to the larger number of galaxies estimated to have unperturbed SF (i.e. with $t_{gal} < t_{AGN}$). To perform a more complete analysis, we also show the results for a combined spectral model obtained by summing up two coeval stellar populations: a single burst model to a quenched SF assuming a variable or a fixed mass ratio of the two populations. This model takes into account the coexistence of stellar populations having different properties: some stars formed in a single initial burst (SSp); in addition to this component, there is

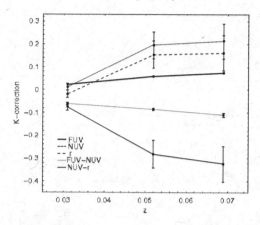

Figure 12.19 Median K-corrections derived by fitted synthetic spectra for bands FUV, NUV, r and colour $FUV - NUV$ and $NUV - r$. The bars are the median scatters.

gas that cools to form stars continuously and is affected by AGN. What we see in Figure 12.18 is that these combined models give the same results as using the single component ones. Our estimates of $t_{gal} - t_{AGN}$ are robust, since the bulk of their distribution depends little on the duration of the unperturbed SF and Z. Despite this result, as we will see later, the past SF history of galaxies is dependent on the choice of these models. Finally, two shortcomings have to be discussed. If we consider more protracted unperturbed SF, we will obtain a larger number of galaxies with higher $t_{gal} - t_{AGN}$ values than using SFs with lower τ (see Figure 12.18). In addition, a similar effect is given if quenching is not almost instantaneous as in our simulation; also in this case we would obtain larger values of $t_{gal} - t_{AGN}$. However, our fast quenching is a good approximation to describe the effect of AGNs for a wide sample of galaxies, but further analysis on these aspects waits to be done.

In order to avoid confusion, we will use in the following the results obtained in the most general case where Z, τ, t_{gal} and t_{AGN} are all unconstrained. Also, K-corrections deserve particular attention. In fact, due to uncertainties in the UV spectrum, it can be difficult to quantify the size of this correction, a circumstance that can introduce systematics. In our range of redshifts, we find typical NUV corrections of 0–0.3, in good agreement with those found in Kaviraj *et al.* (2007) using the same galaxy catalogue. In Figure 12.19, we show results for K-corrections of FUV, r and colours $FUV - NUV$ and $NUV - r$. We obtain a sharp rise in NUV correction: this is negligible in the lowest redshift bin, while its median value becomes ~ 0.25 at higher z. The median scatter in the estimated values of NUV and FUV corrections is consistent with typical uncertainties of observed magnitudes in these bands. Note that Kaviraj *et al.* (2007), using a 9 Gyr old ssp and

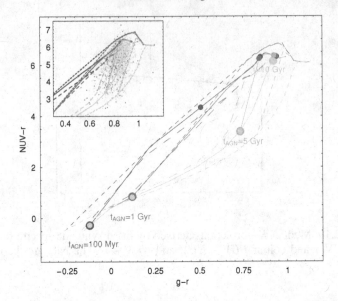

Figure 12.20 Colour–magnitude diagrams of our sample, colour-coded following the classification shown in the legend: $t_{gal} - t_{AGN} < 0$ (orange), $0 \leq t_{gal} - t_{AGN} < 0.5$ (green), $0.5 \leq t_{gal} - t_{AGN} < 1$ (blue), $1 \leq t_{gal} - t_{AGN} < 4$ (violet), $4 \leq t_{gal} - t_{AGN}$ (red). In the inset we show $z = 0$ colours predicted using a $z_f = 3$ single burst population for $Z = Z_\odot$ as a continuous grey line and various $Z < Z_\odot$ (i.e. $Z = 0.0001, 0.0004, 0.004, 0.008$) as dashed grey lines. *Main figure.* We plot NUV-r vs r magnitude. *A colour version of this figure is available in (Tortora et al. 2009).*

Rawle *et al.* (2008), using both SSP and frosting models, find larger corrections of 0.2–1.2 for galaxies having $z \sim 0.1$.

12.3.9 *Quenching event*

In Figure 12.20 we show the visible and UV colour–magnitude diagrams for our sample. As already detailed in Kaviraj *et al.* (2007), the UV diagram in the right panel of this figure is largely different from similar colour diagrams for optical bands (see left panels), since it shows a larger scatter. Figure 12.20 shows the power of $NUV - r$ colour in selecting galaxies with different SFR histories. In particular, the points in the figure are colour-coded according to the fitted value of $t_{AGN} - t_{gal}$: galaxies that experienced some quenching phenomena in the past lie higher in this diagram, while those having lower values of this quantity populate on average a region with a flatter UV continuum and a larger scatter.

A rough comparison with the expectation of the monolithic scenario has been made, superimposing the predicted colours for single burst populations with a more extended formation redshift starting at $z_f = 3$. For completeness, in addition to a model with $Z = Z_\odot$ we also show the predictions for various metallicities $< Z_\odot$ implemented within the Bruzual–Charlot prescription. These predictions are not very sensitive to the precise value of z_f (at least when considering relatively old stellar populations), and our choice is consistent with optical analyses that estimate $z_f \gtrsim 2$ (Bower *et al.* 1992). We see that an almost solar SSP model is only consistent with redder galaxies for both $g - r$ and $NUV - r$ colours. Thus, while galaxies with low $g - r$ and $NUV - r$ can be matched by models with low Z, galaxies in the $u - r$ vs. r plot appear to depart from these predictions that systematically lead to redder colours. However, as we pointed out in a previous subsection, luminous ETGs are almost solar and very low values of metallicities like those used in Figure 12.20 are not applicable to our galaxies (Panter *et al.* 2008; Tortora *et al.* 2009). Moreover, semi-analytic simulations predict that a single solar metallicity is a reasonable hypothesis to describe ETGs (Nagashima and Okamoto 2006), and similar results for the brighter and more massive galaxies have been obtained in N-body + hydrodynamical simulations (Romeo *et al.* 2008). Thus, the comparison with a solar metallicity SSP indicates that only a few galaxies (6–7%) are consistent (within the errors) with a single and old stellar population; these galaxies are the redder ones mainly with $t_{gal} - t_{AGN} \gtrsim 1$. This result is consistent with the $\sim 1\%$ of purely passive galaxies found in Kaviraj *et al.* (2008), analysing a sample of galaxies at $z \gtrsim 0.5$, since in the local universe we observe a larger number of quiescent and passive galaxies.

In Figure 12.21 we show $NUV - r$ vs. $g - r$, with points colour-coded according to Figure 12.20. A correlation is observed, but note that the excursion in $NUV - r$ is much larger than in $g - r$.

Galaxies with a stronger ultraviolet flux are on average characterized by higher RSF, that translates into lower values of $t_{gal} - t_{AGN}$, while those with a lower NUV flux have SF that is quenched early. This correlation is shown in Figure 12.22. In the inner panel, we show the trend with colour $NUV - r$, implicit in Figure 12.20.

The results concerning the scale of SF are also interesting. When considering the results obtained by fitting our reference spectral library, we find that only $\sim 15\%$ of galaxies have $\tau = 1$ Gyr, while a more protracted SF is recovered for the other galaxies in the sample, with $\sim 43\%$ and $\sim 42\%$ of these having $\tau = 3$ and 5 Gyr. As a comparison, we also fitted unperturbed SFs with $\tau \in (0.1, 5)$ Gyr and $Z \in (0.008, 0.02, 0.05)$ obtaining that $\sim 86\%$ of galaxies have $\tau \leq 1$, while less than 1% have $\tau > 3$ Gyr (see also (Tortora *et al.* 2009)). Thus, the low SF scales recovered when simple unperturbed exponential SFs[4] are fitted to data can hide a galaxy population with a more protracted background SF that is quenched in the

---

[4] Similar considerations can be made if we assume a delayed SF.

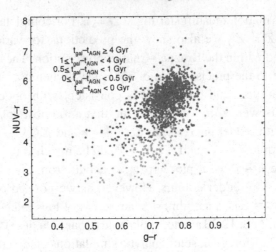

Figure 12.21 UV/optical colour–colour diagram. We plot $NUV - r$ vs. $g - r$ of galaxies in our sample colour-coded following the classification shown in the legend. For colour version see http//:www.cambridge.org/9780521192545.

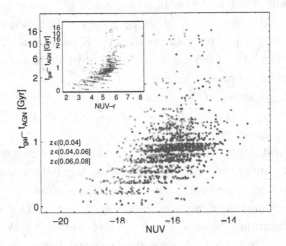

Figure 12.22 $t_{gal} - t_{AGN}$ as a function of ultraviolet magnitude NUV, while in the inserted panel is plotted $t_{gal} - t_{AGN}$ as a function of colour $NUV - r$. Points are colour coded for various redshift bins. See Figure 12.18 for details on the scale used on the y-axes. For colour version see http//:www.cambridge.org/9780521192545.

late stage of galaxy evolution for a feedback effect. These results are consistent with the general scenario depicted for the colour evolution of E + A galaxies in Kaviraj *et al.* (2007). They find that, superimposed on an early burst of formation, a recent SF burst (which typically takes place within a Gyr) over a timescale ranging between 0.01 and 0.2 Gyr is needed to match galaxy colours. These galaxies are just migrating towards the red sequence and show an SF quenching that is correlated with their stellar mass and velocity dispersion and linked to different sources of feedback (AGN or supernovae). However, although different models can have the same final result as our unperturbed and truncated SFs for different class of galaxies, this work reproduces typical timescales consistent with those discussed here (see also Figure 12.13).

Fitting our reference library, we find that $\sim 35\%$ of galaxies have $t_{gal} > t_{univ}$, while this percentage decreases to $\sim 15\%$ for those galaxies with $t_{gal} > t_{univ} + 2$. Excluding all galaxies with $t_{gal} > t_{univ}$ we find a median formation redshift of $z_f = 1.0^{+1.4}_{-0.4}$ (uncertainties are 25th and 75th percentiles); note that the distribution has a strong tail for high formation redshift. For t_{AGN} we find the best estimate $z_{AGN} = 0.13 \pm 0.02$.

12.3.10 Stellar mass fraction and AGN feedback

Until now, we have discussed how the quenching event is linked to colours and galaxy luminosities, but we have not expressly discussed the amount of stellar mass that is produced. To quantify this recent star formation, we define the RSF via the amount of stars produced (i.e. the produced stellar mass fraction) in the last 1 Gyr in the rest frame of the galaxy (Kaviraj *et al.* 2007, 2008). As discussed in the previous subsections, this quantity is strictly dependent on the time of the quenching event, and in particular on $t_{gal} - t_{AGN}$. The latter has been shown to depend on colour $NUV - r$ and the luminosity in the NUV band, while it is less sensitive to the same quantities obtained using only visible bands.

Galaxies with brighter ultraviolet fluxes and bluer $NUV - r$ colours have much more RSF with respect to fainter ones. This result is not surprising, since an indication of it has been obtained by the trends shown in Figure 12.22. In Figure 12.23 we show the RSF as a function of NUV and $NUV - r$. In the left panel, we show the results using our reference library. Obviously, only galaxies with $t_{gal} - t_{AGN}$ less than 1 Gyr survive in these plots and show the presence of RSF. In this case, the trends are not so tight, since some galaxies depart from them (red points). As a comparison, in the right panel we show the results for a model with fixed values of Z and τ, where the correlations are tighter. Coming back to the results we obtained using our reference library, we show in Figure 12.24 the results already plotted in the left panel of Fig. 12.23, but now coded for metallicity and τ, respectively in the left and right panels. These plots show that those galaxies which experience a significant RSF also have metallicities significantly different from the average. In fact, galaxies with different metallicities seem to stay on three different sequences, with a lot of supersolar galaxies having red colours and a low

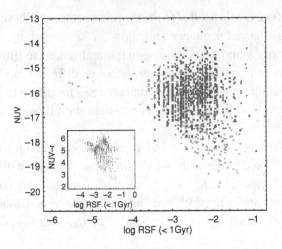

Figure 12.23 NUV (in the main figure) and (NUV-r) (in the inset) vs. RSF for a model leaving free all parameters. The colour code is the same as in Figs. 12.20 and 12.21. *A colour version of this figure is available in (Tortora et al. 2009).*

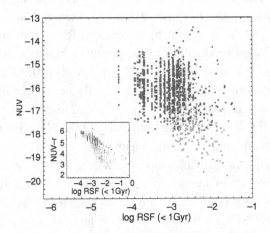

Figure 12.24 NUV (in the main figure) and (NUV-r) (in the inset) vs RSF for a model leaving free all parameters. In the inset panel we colour for Z, with $Z = 0.008$ (green), $Z = 0.02$ (blue) and $Z = 0.05$ (red). In the main Figure we colour for τ with $\tau = 1\,\text{Gyr}$ (red), $\tau = 3\,\text{Gyr}$ (blue) and $\tau = 5\,\text{Gyr}$ (green). *A colour version of this figure is available in (Tortora et al. 2009).*

NUV flux, but a large SF. This trend is more evident if we look at the plot as a function of colour. The trends with SF timescale τ are less clear, but some of those galaxies departing from the mean trend have $\tau = 5\,\text{Gyr}$. Thus, some systematics might be introduced into these results by the degeneracies between Z and τ, and possibly with other parameters. However, our results are robust, since qualitatively we recover some trends that do not depend on the particular choice of spectral library.

For the model that leaves all of the parameters free, we find a median RSF of $\sim 0.2^{+0.4}_{-0.2}\%$. In particular, galaxies with the bluest colours (and with larger NUV fluxes) have an RSF of $\sim 1\%$, while the redder ones have an RSF less than 0.1%. On the contrary, a lower median RSF of $\sim 0.1\%$ is observed for the model with $\tau = 3\,\mathrm{Gyr}$ and $Z = Z_\odot$.

On average, our recovered RSF fractions are slightly lower than those obtained in Kaviraj *et al.* (2007), but this is not surprising since we follow a different approach. In detail, we adopt what we can call a *blue-to-red* approach, since we have modelled galaxies using an exponential SF. Thus, a galaxy is initially blue, and later it becomes red since SF is quenched. On the contrary, using a *red-to-blue* approach, galaxies are modelled with a single burst population, which predicts systematically redder colours than those obtained with a more complex SF. Thus, to match the observations one needs a recent burst to make the colours bluer. It is, however, worth noting that within scatter the two analyses give fully compatible results.

Our results can also be connected with the downsizing scenario. In fact, the fact that redder galaxies (i.e. with a lower NUV flux) have less RSF with respect to the bluer ones can reproduce the trends found in recent works (e.g. Romeo *et al.* 2008; Tortora *et al.* 2009). However, this connection has to be better analysed and further information could be recovered by extending the luminosity range.

12.3.11 SED evolution

In the previous subsection, we obtained results that indicate the presence of an RSF that is connected with the quenching events. Here, we quantify the evolutive path of galaxies in our sample, discussing how colours and magnitudes (averaged over the galaxy sample) evolve. A more detailed analysis of these results is beyond the scope of this chapter.

We have verified that a large number of the galaxies in our sample are not very sensitive to changes in the details of the spectral library, since they predict a median $t_{\mathrm{gal}} - t_{\mathrm{AGN}}$ that does not change siginficatly (see Figure 12.18). Thus, our models are able to successfully describe the recent formation history of galaxies, while the extrapolation to early phases of galaxy evolution is influenced by the choice of unperturbed synthetic spectra. However, this extrapolation can fail to describe the earlier evolution, since galaxies that have experienced various (minor or major) mergers and AGNs can act to quench SF at different epochs (Khalatyan *et al.* 2008). Therefore, both SF, magnitude and colour evolution can be very complex and it is not possible to probe them efficiently.

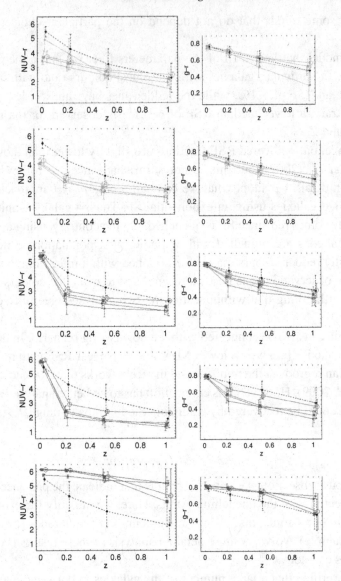

Figure 12.25 Colour evolution predicted by fit. The different panels refer to values of $t_{gal} - t_{AGN}$ according to the colour coding of Figures 12.20, 12.21 and 12.23. We show the results from our reference model leaving all parameters free to change (star symbols and short-dashed error bars), $Z = Z_\odot$ and $\tau = 3\,\mathrm{Gyr}$ (diamond symbols and continuous error bars), two combined models with a $Z = Z_\odot$ burst superimposed on a truncated exponential SF with $\tau = 3\,\mathrm{Gyr}$ with a free mass ratio (box symbols and long-dashed error bars) and one with single burst amounting to the 75% of the total mass (circle symbols and thin continuous error bars). The black symbols and bars are obtained using an unperturbed SF with τ and Z free to change (averaged over all galaxies in the sample). At each redshift, the different models are artificially shifted to make the appearance clearer. *Left panels*: $NUV - r$ vs. z. *Right panels*: $g - r$ vs. z.

We map the evolution of rest-frame galaxy colours up to 7–8 Gyr ago, determined from our best-fit models for three values of the look-back time t_{lbt} corresponding to redshift $z = 0.2, 0.5, 1$. In Figure 12.25 we show the median values and median deviations for $NUV - r$ and $g - r$ of galaxies in the sample. Actually, colour and luminosity evolution is stronger for galaxies with a more recent event of SF and quenching (i.e. lower $t_{gal} - t_{AGN}$). These galaxies became red only recently, while those with a quenching event that occurred many Gyr ago were already on the red sequence. While changes of 2–3 magnitudes in $NUV - r$ are observed, $g - r$ changes are ~ 1 magnitude. In addition to our reference model, we show the evolution of colours obtained using a library of truncated SFs with $Z = Z_\odot$ and $\tau = 3$ Gyr and two mixed models consisting of a solar single burst and a truncated SF with $Z = Z_\odot$ and $\tau = 3$ Gyr, leaving free a proportion of the two populations, and fixing the single burst to 75% of the total mass. These latter two models leave the galaxy colours for $t_{gal} > t_{AGN}$ almost unchanged, while affecting the early history of galaxies producing redder colours, this effect depending on the ratio of the two populations. For comparison, we show the median colours obtained by fitting an unperturbed SF leaving t_{gal}, Z and τ free to change.

These predicted tracks show an interesting result concerning the strength of quenching, which is mainly evident in the $NUV - r$ colour. With the exception of galaxies with $t_{gal} < t_{AGN}$ and those with an early quenching event, the systems with an observed larger colour difference (and thus on average large values of $t_{gal} - t_{AGN}$) were bluer and more star forming at $z \sim 0.2$. Thus, this result seems to indicate that in galaxies with a large SF at $z \gtrsim 0.2$, the effect of AGN quenching can be stronger, to produce the reddest galaxies we observe today. Nonetheless, recent X-ray and optical selected analysis of high redshift AGNs (Martin *et al.* 2007; Nandra *et al.* 2007; Silverman *et al.* 2008a, b) exhibit a large fraction of AGN in galaxies at intermediate/blue colours, which could be driven by AGN feedback toward the red sequence at $z \sim 0$.

The NUV fluxes are systematically larger (corresponding to lower $NUV - r$) than those estimated using unperturbed SFs. On the contrary, the luminosity in the r band is very little dependent on a change in models and the predictions agree quite well. This is not surprising, since redder bands are less sensitive to recent or past SF.

12.4 Conclusions

In this work, we have attempted to build up a realistic model of AGN feedback. We performed a hydrodynamical simulation to analyse the impact of jets produced by AGNs on an inhomogeneous medium, properly representing cold gas clouds that form stars in galaxies. In our simulation, which extends and generalizes a set that we previously performed (Antonuccio-Delogu and Silk 2008), a powerful jet, $P_j = 10^{46}\,\mathrm{erg\,s}^{-1}$, propagates within an inhomogeneous, two-phase ISM containing a realistic distribution of star-forming clouds. SFRs in clouds are described using the empirical SK prescription, where the SFR depends on the mass density of the star-forming regions of the clouds. In an early phase, the shocks advancing before the expanding cocoon tend to compress the cold clouds, without significantly changing the temperature of the medium, thus increasing the SF. Later, when the cocoon has propagated within the medium, the temperature of both the medium and the clouds increases significantly and the mass of clouds is reduced, also due to KH instabilities. Thus, at the beginning, a positive feedback increases SF, but the dominant effect is the negative feedback that quenches SF in a time of $\sim$2–3 $\times$ 10^6 yr. One interesting result of this work is that for the first time a hydrodynamical simulation allows the determination of the effect of jets emitted by AGN on SF in galaxies. Previous work has relied on empirical prescriptions to take into account AGN feedback (Granato *et al.* 2001, 2004; Cattaneo *et al.* 2006; Martin *et al.* 2007), and is thus more robustly supported by our results.

The high jet power in this simulation is probably the main reason for the fact that, at the end of the simulation, all the gas within the computational volume is in a physical state for which all SF is quenched. For lower injection powers, the cocoon expansion will be halted sooner, and the region where SF is quenched will consequently be smaller.

Based on the results of this simulation, we develop a more general model that assumes an SFR for ETGs composed of a background SF with a more or less extended duration, which is quenched by a feedback effect such as that analysed in the first part of this chapter. However, we stress here that this general prescription can also describe the effect of other quenching mechanisms, such as merging, harassment, etc. Restricting ourselves to the case of a high-power jet, as in the present paper, the typical timescales of quenching are very small, and no other sources of feedback can stop SF within such a short time. Note also that galaxy merging seems to be one of the mechanisms that activate AGNs, which after the burst induced by the merger are able to quench the residual SF in the remnant (Di Matteo *et al.* 2005, Khalatyan *et al.* 2008).

Our paradigm is that luminous ETGs are an exception within the zoo of galaxies in the universe. We suppose that the normality in galaxy samples is represented by fainter ETGs, which have a protracted SF, but due to feedback this is quenched and colours redden. This is the main cause of the observed low values of duration of SF in comparison with observations (Tortora *et al.* 2009). On the contrary, semi-analytical galaxy simulations predict more protracted SFR (De Lucia *et al.*

2006): supernovae feedback was taken into account in these simulations, but the disagreement with other observations can be due to an improperly accounted for contribution from AGNs. Springel *et al.* (2005a) simulated the effect of AGNs in the merging of two massive and gas-rich spiral galaxies, obtaining similar results. They found that SF is inhibited with respect to the case without black holes; interestingly, without AGNs this kind of merger is not able to produce red ETGs, which remain blue (with residual SF) even after several Gyr. On the contrary, the presence of black holes reddens galaxy colours much faster, giving $u - r \sim 2.3$ in less than 1 Gyr after the beginning of the merging process.

We assume an unperturbed SF law with a scale $\tau = 1, 3, 5$ Gyr, which is quenched at t_{AGN} leaving us free to change these parameters as well as galaxy ages and metallicities. We fit these models to observations from a cross-matched SDSS+GALEX catalog. Confirming the results in Kaviraj *et al.* (2007), the UV has been shown to be a strong indicator of SF, in particular the UV/optical colour $NUV - r$ allows us to select galaxies with different levels of SF, either stopped in the last Gyr or still in action. We show that the quantity $t_{gal} - t_{AGN}$ is able to describe the physical state of galaxies, indicating how long ago SF stopped. The largest number of galaxies have $t_{gal} - t_{AGN} \sim 0.5$–1 Gyr, indicating the necessity of a RSF phenomenon (up to 1–2 Gyr ago), which is quenched by AGNs. Galaxies with larger $NUV - r$ have higher values of $t_{gal} - t_{AGN}$, i.e. SF is quenched early in their history, while lower values of $NUV - r$ correspond to galaxies with a more recently quenched SF. Finally, galaxies with SF not affected by AGNs have the flattest $NUV - r$ colours. This distinction is not so clear (or absent) for visible colours. These results are shown in Figures 12.20 and 12.21. The epoch of the quenching event appears to be correlated with ultraviolet flux: galaxies bluer and brighter in the NUV band are also those with a more recent (or absent) quenching event. As shown in Figure 12.18, $t_{gal} - t_{AGN}$ is found to depend negligibly on the details of background spectral library (e.g. τ and Z), at least for the bulk of galaxies: thus we obtain a robust estimate of the quenching time and RSF. Our analysis is less sensitive to the earliest phases of galaxy history, notwithstanding the mean evolution of galaxy population analysed here, as shown in Figure 12.25. Finally, a shortcoming of this analysis is related to the timescale of the quenching event, since softer power jets would inhibit SF within a larger timescale. Such a slower quenching model would be fitted to the colours by a larger $t_{gal} - t_{AGN}$, thus our results for this parameter have to be interpreted as lower limits.

One of the most significant results we have found is that the $NUV - g$ colour index is very sensitive to the presence of a very young stellar population. Typical enhancement of 2–3 magnitudes of $NUV - g$ are observed relative to the no-feedback case. Our findings agree with some current simulations that invoke two different modes of AGN feedback. A so-called 'quasar mode' assumes that, during a major merger event at high redshift, a fraction of the gas accreted by a central black hole is injected into the gas of the host galaxy, quenching SF (Springel *et al.* 2005a,b; Di Matteo *et al.* 2005). At later times, another effect is important, i.e. the 'radio mode', which is responsible for making galaxies quiescent, and this effect is driven by low-level AGNs (Croton *et al.* 2006, Bower *et al.* 2006). More recently, Schawinski *et al.* (2009) speak of a 'truncation mode' to indicate AGN feedback at high redshift. At recent epochs, no such strong activities or powerful radio jets have been observed, and so this kind of AGN feedback is often referred to as the 'suppression mode'. These authors suggest that this process could leave a residual SF (e.g. Schawinski *et al.* 2006), but is able to move galaxies along the red sequence.

Further refinements of our analysis are needed. Firstly, it is important to enlarge the sample of galaxies and analyse a more extended range of magnitudes. The galaxies under analysis, due to constraints on magnitude, are relatively bright with $M_B \sim -19$. Different kinds of analysis seem to indicate that brighter and more massive ETGs ($M_B \sim -20.5$ and $M_\star \gtrsim 10^{11} M_\odot$) are fundamentally different from fainter and less massive ones ($M_B \gtrsim -20.5$ and $M_\star \sim 10^{11} M_\odot$). In these two different luminosity and mass regimes, the size–luminosity or size–mass (Shen *et al.* 2003) and Faber–Jackson relation (Matkovic and Guzman 2005) have different slopes. Also, the Sersic index is changes with luminosity (Prugniel and Simien 1997), and dark matter is shown to have bivariate behaviour in the two ranges (Tortora *et al.* 2009), suggesting that physical phenomena allowing the formation of galaxies of different mass and luminosity can be manifold. Thus, a future analysis should also be directed at studying a wider range of luminosities and masses to map the transition from the blue cloud to the red sequence. Also, to connect our results with the downsizing scenario, we need to enlarge the sample to fainter magnitudes, since strong changes in observable quantities such as galaxy age, t_{AGN}, τ and Z are probably visible in these luminosity regimes. We are planning to do other simulations changing the main input parameters, in order to have a more reasonable model of AGN feedback that depends on the jet power, for instance, linking it to the main galaxy observables (Liu *et al.* 2006). Comparison with simulations will allow us to quantify how much AGN feedback has to be implemented, to make both semi-analytical and hydrodynamical simulations able to correctly predict the main properties of galaxies we observe. Such a more complex model of quenching would be parameterized as a function of the main parameters of the system and compared to other AGN feedback prescriptions (Granato *et al.* 2001, Cattaneo *et al.* 2006, Martin *et al.* 2007).

Linking observations at low redshift, such as those analysed here, with high redshift data (Martin *et al.* 2007; Nandra *et al.* 2007; Silverman *et al.* 2008a,b) could certainly be a powerful way to shed light on the galaxy evolution scenario, and in particular on the role that AGNs have in the early and later phases of the SF history of galaxies.

References

Antonuccio-Delogu, V., & Silk, J. (2008). *MNRAS*, **389**, 1750

Baek, C. H., Kang, H., Kim, J., Ryu, D. (2005). *ApJ*, **630**, 689

Baldwin, J. A., Phillips, M. M., & Terlevich, R. (1981). *PASP*, **93**, 5

Blundell, K. M., Rawlings, S., Willott, C. J., Kassim, N. E., & Perley, R. (2002). *New Astronomy Review*, **46**, 75

Bodo, G., Massaglia, S., Ferrari, A., & Trussoni, E. (1994). *A&A*, **283**, 655

Bonnor, W. B. (1956). *MNRAS*, **116**, 351

Bouquet, S., Romain, T., & Chieze, J. P. (2000). *ApJS*, **127**, 245

Bower R. G., Lucey, J. R., & Ellis, R. (1992). *MNRAS*, **254**, 589

Bower, R. G., Benson, A. J., Malbon, R., *et al.* (2006). *MNRAS*, **370**, 645

Bromley, J. M., Somerville, R. S., & Fabian, A. C. (2004). *MNRAS*, **350**, 456

Bruzual, G., & Charlot, S. (2003). *MNRAS*, **344**, 1000

Cattaneo, A., Dekel, A., Devriendt, J., Guiderdoni, B., & Blaizot, J. (2006). *MNRAS*, **370**, 1651

Combes, F., Young, L. M., & Bureau, M. (2007). *MNRAS*, **377**, 1795

Croton, D. J. Springel, V. White, S. D. M., *et al.* (2006). *MNRAS*, **365**, 11

De Lucia, G., Springel, V., White, S. D. M., Croton, D., & Kauffmann, G. (2006). *MNRAS*, **366**, 499

Di Matteo, T., Springel, V. & Hernquist, L. (2005). *Nature*, **433**, 604

Ebert, R. (1955). *Zeitschrift fur Astrophysik*, **37**, 217

Fryxell, B., *et al.* (2000). *ApJS*, **131**, 273

Goudfrooij, P., Hansen, L., Jorgensen, H. E., & Norgaard-Nielsen, H. U. (1994). *A&AS*, **105**, 341

Granato, G. L., Silva, L., Monaco, P., *et al.* (2001). *MNRAS*, **324**, 757

Granato, G. L., De Zotti, G., Silva, L., Bressan, A., & Danese, L. (2004). *ApJ*, **600**, 580

Haehnelt, M. G. (2004). *Coevolution of Black Holes and Galaxies*, Cambridge University Press, p. 405

Hester, J. A. (2006). *ApJ*, **647**, 910

Iliev, I. T., & Shapiro, P. R. (2001). *MNRAS*, **325**, 468

Inoue, S., & Sasaki, S. (2001). *ApJ*, **562**, 618

Kaiser, C. R., & Alexander, P. (1997). *MNRAS*, **286**, 215

Kauffmann, G., *et al.* (2003). *MNRAS*, **346**, 1055

Kaviraj, S., Kirkby, L. A., Silk, J., & Sarzi, M. (2007). *MNRAS*, **382**, 960

Kaviraj, S., *et al.* (2008). *MNRAS*, **388**, 67

Kennicutt, R. C., Jr. (1998). *ApJ*, **498**, 541

Khalatyan, A., Cattaneo, A., Schramm, M., *et al.* (2008). *MNRAS*, **387**, 13

Klein, R. I., McKee, C. F., & Colella, P. (1994). *ApJ*, **420**, 213

Krause, M., & Alexander, P. (2007). *MNRAS*, **376**, 465

Kritsuk, A. G., Padoan, P., Wagner, R., & Norman, M. L. (2007). *Turbulence and Nonlinear Processes in Astrophysical Plasmas*, **932**, 393

Liu, Y., Jiang, D. R., & Gu, M. F. (2006). *ApJ*, **637**, 669

Macchetto, F., Pastoriza, M., Caon, N., *et al.* (1996). *A&AS*, **120**, 463

Mahmood, A., Devriendt, J. E. G., & Silk, J. (2005). *MNRAS*, **359**, 1363

Martin, D. C., GALEX collaboration (2005). *ApJ*, **619**, L1

Martin, D. C., *et al.* (2007). *ApJS*, **173**, 342

Matkovic, A., & Guzman, R. (2005). *MNRAS*, **362**, 289

Mellema, G., Kurk, J. D., & Röttgering, H. J. A. (2002). *A&A*, **395**, L13

Morganti, R., *et al.* (2006). *MNRAS*, **371**, 157

Nagashima, M., & Okamoto, T. (2006). *ApJ*, **643**, 863

Nakamura, F., McKee, C. F., Klein, R. I., & Fisher, R. T. (2006). *ApJS*, **164**, 477

Nandra, K., *et al.* (2007). *ApJ*, **660**, L11

Padoan, P., Jones, B. J. T., & Nordlund, A. P. (1997). *ApJ*, **474**, 730

Panter, B., Jimenez, R., Heavens, A. F., & Charlot, S. (2008). *MNRAS*, **391**, 1117

Passot, T., & Vázquez-Semadeni, E. (1998). *Phys. Rev. E*, **58**, 4501

Perucho, M., Hanasz, M., Martí, J. M., & Sol, H. (2004). *A&A*, **427**, 415

Perucho, M., Martí, J. M., & Hanasz, M. (2005). *A&A*, **443**, 863

Priest, E. R. (1987). *Solar Magneto-hydrodynamics.* Dordrecht: D. Reidel

Prugniel, Ph., & Simien, F. (1997). *A&A*, **321**, 111

Rawle, T. D., Smith, R. J., Lucey, J. R., Hudson, M. J., Wegner, G. A. (2008). *MNRAS*, **385**, 2097

Romeo, A. D., Napolitano, N. R., Covone, G., *et al.* (2008). *MNRAS*, **389**, 13

Salim, S. (2007). *ApJS*, **173**, 267

Sarzi, M., *et al.* (2008). *Pathways Through an Eclectic Universe*, ASP Conference Series, **390**, 218

Schawinski, K., *et al.* (2006). *Nature*, **442**, 888

Schawinski, K., *et al.* (2009). *ApJ*, **690**, 1672

Schaye, J., & Dalla Vecchia, C. (2008). *MNRAS*, **383**, 1210

Schmidt, M. (1959). *ApJ*, **129**, 243

Schmidt, M. (1963). *ApJ*, **137**, 758

Shabala, S. S., Ash, S., Alexander, P., & Riley, J. M. (2008). *MNRAS*, **388**, 625

Shapiro, P. R., Iliev, I. T., & Raga, A. C. (1999). *MNRAS*, **307**, 203

Shen, S., Mo, H. J., White, S. D. M., *et al.* (2003). *MNRAS*, **343**, 978

Silk, J., & Rees, M. J. (1998). *MNRAS*, **331**, L1

Silverman, J. D., Mainieri, V. Lehmer, B. D. *et al.* (2008). *ApJ*, **675**, 1025

Silverman, J. D. *et al.* 2008, arXiv:0810.3653

Springel, V., Di Matteo, T., & Hernquist, L. (2005). *ApJ*, **620**, 79

Springel, V., Di Matteo, T., & Hernquist, L. (2005). *MNRAS*, **361**, 776

Sutherland, R. S., & Bicknell, G. V. (2007). *ApJS*, **173**, 37

Sutherland, R. S., & Dopita, M. A. (1993). *ApJS*, **88**, 253

Tasker, E. J., & Bryan, G. L. (2006). *ApJ*, **641**, 878

Thomas, D., Maraston, C., Schawinski, K., *et al.* (2007). *IAU Symposium*, **241**, 546

Tortora, C., Antonuccio-Delogu, V., Kaviraj, S., *et al.* (2009). *MNRAS*, **396**, 61

Vázquez-Semadeni, E. (1994). *ApJ*, **423**, 681

Wada, K., & Norman, C. A. (2001). *ApJ*, **547**, 172

Wada, K., & Norman, C. A. (2007). *ApJ*, **660**, 276

Yi, S. K., Yoon, S.-J., Kaviraj, S., Deharveng, J.-M., and the GALEX Science Team. (2005). *ApJ* **619**, L111.

Yi, S. K., Kaviraj, S., & Schawinski, K. (2007). *Revista Mexicana de Astronomía y Astrofísica Conference Series*, **28**, 109

13

Large-scale expansion of AGN outflows in a cosmological volume

P. Barai

13.1 Introduction

Outflows from AGN are observed in a wide variety of forms: radio galaxies, broad absorption line quasars, Seyfert galaxies exhibiting intrinsic absorption in the UV, broad emission lines, warm absorbers and absorption lines in X-rays (e.g. Creenshaw *et al.* 2003; Everett 2007). There have been studies on the cosmological impact of quasar outflows on large scales (Furlanetto & Loeb 2001, hereafter FL01; Scannapieco and Oh 2004, hereafter SO04; Levine & Gnedin 2005, hereafter LG05). Barai (2008) investigated the cosmological influence of radio galaxies over the Hubble time. All these studies considered spherically expanding outflows.

On cosmological scales an outflow is expected to move away from the high-density regions of large-scale structures, with the outflowing matter getting channelled into low-density regions of the universe (Martel and Shapiro 2001). We implement such anisotropic AGN outflows within a cosmological volume. The simulation methodology is given in Section 13.2, and the results are discussed in Section 13.3.

13.2 The numerical setup

13.2.1 N-body simulation and distribution

We simulate the growth of large-scale structures in a cubic cosmological volume of comoving size $L_{box} = 128h^{-1}$ Mpc. We use the particle-mesh (PM) algorithm, with 256^3 equal-mass particles, on a 512^3 grid. A particle has a mass of $1.32 \times 10^{10} M_\odot$, and the grid spacing is $\Delta = 0.25h^{-1}$ Mpc. We consider a concordance

AGN Feedback in Galaxy Formation, eds. V. Antonuccio-Delogu and J. Silk. Published by Cambridge University Press. © Cambridge University Press 2011.

ΛCDM model with the cosmological parameters: $\Omega_M = 0.268$, $\Omega_\Lambda = 0.732$, $H_0 = 70.4\,\text{km}\,\text{s}^{-1}\text{Mpc}^{-1}$, $\Omega_b = 0.0441$, $n_s = 0.947$, and $T_{CMB} = 2.725$.

The redshift-dependent luminosity distribution of AGN is obtained from the bolometric quasar luminosity function (QLF) (Hopkins *et al.* 2007),

$$\phi(L, z) \equiv \frac{d\Phi}{d\log L} = \frac{\phi_\star}{(L/L_\star)^{\gamma_1} + (L/L_\star)^{\gamma_2}}, \tag{13.1}$$

which gives the number of quasars per unit comoving volume, per unit $\log_{10}$ of luminosity. A fraction $f_{\text{outflow}} = 0.2$ of AGN are considered to host outflows (Ganguly and Brotherton 2008). The number of outflows within the simulation box of comoving volume $V_{\text{box}} = L_{\text{box}}^3$, at epoch z, between $[L, L+dL]$ is

$$N(L, z) = f_{\text{outflow}}\,\phi(L, z)\,\text{d}[\log_{10} L]V_{\text{box}}. \tag{13.2}$$

The AGN activity lifetime is taken as $T_{\text{AGN}} = 10^8$ yr; the minimum and maximum AGN luminosities as $10^8 L_\odot$ and $10^{14} L_\odot$.

Using the QLF, we obtain the entire cosmological population of AGN in the simulation volume starting from $z = 6$, namely the birth redshift (z_{bir}), switch-off redshift (z_{off}) and bolometric luminosity (L_{bol}) of each source. A total of 929 805 sources were produced.

At each timestep, we filter the density distribution on the 512^3 grid (from the PM code) using a Gaussian filter containing a mass $10^{10} M_\odot$, assumed as the minimum mass of a halo hosting an AGN. We identify the density peaks, or the grid cells where the filtered density exceeds the values at the 26 neighboring cells. We consider the peaks that have a filtered density $> 5\times$ the mean density of the box, and each new AGN born during that epoch (whose z_{bir} values fall within the timestep interval) is located at the center of one such peak cell, selected randomly. After their initial distribution, the AGNs are allowed to evolve according to the prescription in Subsection 13.2.2.

13.2.2 Outflow model

Despite the observational differences between various outflows, we stress that the AGNs hosting outflows constitute a random subset of the whole AGN population, and we simply assume the same outflow model for all AGNs (also FL01, SO04, LG05). We allow each outflow to evolve through an active-AGN life (Subsection 13.2.2), when $z_{\text{bir}} > z > z_{\text{off}}$. After the central engine has stopped activity (when $z < z_{\text{off}}$), it enters the late-expansion phase (Subsection 13.2.2).

The baryonic ambient gas density, $\rho_g(z, \mathbf{r})$, is considered to follow the dark matter density, $\rho_M(z, \mathbf{r})$, in the N-body simulation: $\rho_g = (\Omega_b/\Omega_M)\rho_M$. The external gas pressure is $p_g(z, \mathbf{r}) = \rho_g(z, \mathbf{r})kT_g/\mu$. The external temperature is fixed at $T_g = 10^4$ K, assuming a photoheated ambient medium, and $\mu = 0.611$ amu is the mean molecular mass.

The active life

The AGN activity period is short compared to the Hubble time. So we neglect energy losses and Hubble expansion of the cosmological volume when the quasar is active. We approximate the shape of the expanding outflow as spherical with radius R.

An active outflow is inflated by twin collimated relativistic jets expanding from the central AGN (Begelman *et al.* 1984), each of length R. We consider that the kinetic luminosity carried by each jet is a constant fraction of the AGN bolometric luminosity: $L_K = \epsilon_K L_{bol}/2$, with the kinetic fraction as $\epsilon_K = 0.1$ (FL01; LG05; Shankar *et al.* 2008). The jet advance speed is obtained by balancing the jet momentum flux with the ram pressure of the ambient medium: $L_K/(A_s c) = \rho_g (dR/dt)^2$. Here, A_s is the area of the shocked "working" surface at the jet head. We use $A_s = 2\pi R^2 \theta_s^2$, assuming that the shock front has a constant half-opening angle of $\theta_s = 5°$ relative to the central AGN (FL01). All the kinetic energy transported along the jets during an AGN's age $t_{age} = t(z) - t(z_{bir})$ is transferred to the outflow. The outflow energy is $E_0 = 2L_K t_{age}$, and its pressure follows $p_0 V_0 = (\Gamma_0 - 1)E_0$. The adiabatic index of the relativistic outflow plasma is $\Gamma_0 = 4/3$.

The outflow volume during this active spherical expansion is $V_0(z) = 4\pi R^3/3$.

The late anisotropic expansion

When AGN activity ends, the left-over high-pressure outflow expands into the large scales of the IGM with an anisotropic geometry. It is represented as a "bipolar spherical cone" with radius R and opening angle α (Pieri *et al.* 2007), and it follows the direction of least resistance (DLR). We perform a second-order Taylor expansion of the density around each peak, whose coefficients are determined by performing a least-squares fit to all the grid cells within a distance 2Δ from the peak. We then rotate the coordinate axes such that the cross-terms vanish to give $\rho(x', y', z') = \rho_{peak} - Ax'^2 - By'^2 - Cz'^2$. The largest of the coefficients A, B, C gives the DLR.

The Sedov–Taylor adiabatic blast wave model is used to obtain the radius of the overpressured outflow (Castor *et al.* 1975; SO04; LG05): $R(z) = \xi_0(E_{tot}t_{age}^2/\overline{\rho_g}(z))^{1/5}$. We obtain $\overline{\rho_g}(z)$ by averaging the gas density of the grid cells in the simulation box occurring within the outflow volume. For a strong explosion in the $\Gamma_g = 5/3$ ambient gas, $\xi_0 = 1.12$. Here the total kinetic energy injected into the outflow by the AGN throughout the active lifetime is $E_{tot} = 2L_K T_{AGN}$.

Adiabatic expansion losses are considered, and the outflow pressure evolves as $p_0 R^{3\Gamma_g} = $ constant, with the constant derived from the values at the end of the active phase. The outflow follows an anisotropic expansion as long as its pressure exceeds the external pressure of the IGM, i.e. $p_0(z) > p_g(z)$. During this late biconical expansion the outflow volume is $V_0(z) = 4\pi R^3 (1 - \cos(\alpha/2))/3$.

When $p_0(z) \leq p_g(z)$, or the outflow has reached pressure equilibrium with the external IGM, it has no further expansion. After this, the outflow simply evolves passively with the Hubble flow of the cosmological volume. Thus an outflow in pressure equilibrium attains a final volume of $V_0 = 4\pi R_f^3 (1 - \cos(\alpha/2))/3$, where R_f is the final comoving radius of the outflow.

13.3 Results and discussion

At each time-step the total volume occupied by the AGN outflows is computed by counting the contributions of all the sources born by then, both the active ones and those in the anisotropic phase. We performed four simulations with opening angles of $\alpha = 60°, 90°, 120°, 180°$, all with $\epsilon_K = 0.1$, and one with $\alpha = 120°$ and $\epsilon_K = 0.05$, whose results are shown in the figures.

Figure 13.1 shows the redshift evolution of a single AGN outflow in the simulation. The different phases of the expansion are separated by vertical lines: active-AGN phase on the left, post-AGN phase in the middle, and passive Hubble expansion on the right. At the end of the active phase ($z = 5.31$) the outflow is overpressured by a factor of ~25. So it continues to expand while its pressure falls faster because of adiabatic losses. Finally when p_0 falls to a level to match the external pressure it does not expand anymore. From $z = 4.02$ the comoving outflow radius remains constant at $0.48h^{-1}$ Mpc in the passive Hubble phase.

This illustrates that the AGN outflows are persistently overpressured for a significant period of time even after the central engine has stopped activity, and hence continue to expand into the ambient medium. In Figure 13.1, after an active life of $T_{AGN} = 10^8$ yr, the outflow remains overpressured for $\sim4.6 \times 10^8$ yr. Such results are in accord with other studies (Barai 2008).

We count the grid cells in the simulation box that occur inside the volume of one or more AGN outflows. The total number of these filled cells, N_{AGN}, gives the total volume of the box occupied by outflows. We express the total volume filled as a fraction of volumes of various overdensities in the box, $N_\rho = N(\rho > C\bar{\rho})$, where $\bar{\rho} = (1 + z)^3 \Omega_M 3 H_0^2 / (8\pi G)$ is the mean matter density of a spatially flat universe (the box) at an epoch z. So N_ρ gives the number of cells that are at a density C times the mean density. We find N_ρ for $C = 0, 1, 2, 3, 5, 7$; $C = 0$ gives the total volume of the box.

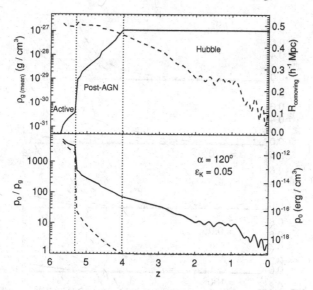

Figure 13.1 Characteristic quantities for the evolution of a single AGN outflow with bolometric luminosity $L_{bol} = 3.5 \times 10^9 L_{\odot}$, opening angle $\alpha = 120°$, and kinetic fraction $\epsilon_K = 0.05$, as a function of redshift. *Upper panel:* comoving radius of outflow (R, *solid*) with y-axis label at top-right, and mean ambient gas density within the outflow volume ($\overline{\rho_g}$, *dashed*) with y-axis label at top-left. *Lower panel:* pressure inside the outflow (p_0, *solid*)) with y-axis label at bottom-right, and overpressure factor of the outflow w.r.t. the external medium (p_0/p_g, *dashed*) with y-axis label at bottom-left. The vertical *dotted* lines separate the different phases of expansion of the outflow: active, post-AGN overpressured, and the final passive Hubble evolution.

Figure 13.2 shows the redshift evolution of the volume filling factors for our five simulation runs. With 10% kinetic efficiency, 0.13 of the entire universe is filled at present by AGN outflows with an opening angle of 60°; the fraction increases to 0.17 with 90°, 0.21 with 120°, and 0.25 with 180°. A 5% kinetic efficiency and $\alpha = 120°$ fills 0.13 of the volume. In all our runs, the outflows fill up all of the regions with $\rho > 2\overline{\rho}$ by $z = 0.3$. With $\epsilon_K = 0.1$ and $\alpha = 90°$ or higher, the outflows permeate all the overdense regions ($\rho > \overline{\rho}$) by $z = 0.1$.

It is the overdense regions of the universe that gravitationally collapse to form stars and galaxies. So evidently the AGN outflows have a profound cosmological impact on the protogalactic regions. We note that the volumes obtained by LG05 (100% filling by $z \sim 1$) are higher than ours.

Blustin and Fabian (2009) used constraints imposed by the observed radio emission and obtained upper limits to the volume filling factors of nearby AGN winds to be in the range $10^{-4} - 0.5$. The volume filling fractions of AGN outflows we obtained in our simulations are well within this observational range.

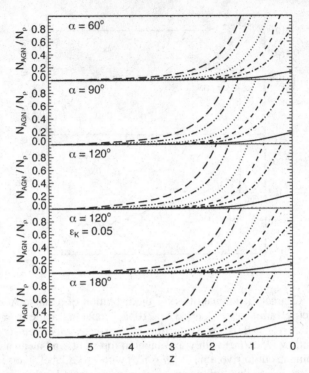

Figure 13.2 Volume of the simulation box filled by AGN outflows (N_{AGN}) as a fraction of the total volume (*solid*), and as a fraction of volumes of various overdensities: $N(\rho > \overline{\rho})$ (*dash dot*), $N(\rho > 2\overline{\rho})$ (*dashed*), $N(\rho > 3\overline{\rho})$ (*dotted*), $N(\rho > 5\overline{\rho})$ (*dash dot dot dot*), $N(\rho > 7\overline{\rho})$ (*long dashes*).

We perform preliminary estimates of the energy density and magnetic field in the volumes of the universe filled by the AGN outflows. The energy density inside the outflow behaves similarly to the outflow pressure evolving adiabatically (Subsection 13.2.2), $u_{\text{E}} = 3p_0$. Assuming equipartition of energy between magnetic field of strength B_0 and relativistic particles inside the outflow, the magnetic energy density is $u_{\text{B}} = u_{\text{E}}/2 = B_0^2/(8\pi)$. We define the volume weighted average of a physical quantity $\mathcal{A}$ as $\langle \mathcal{A} \rangle(z) \equiv \sum(\mathcal{A}V_0)/\sum V_0$, where the summation is over all outflows existing in the simulation box at that epoch.

Figure 13.3 shows the redshift evolution of the total volume filling fraction, $\langle u_{\text{E}} \rangle$ and $\langle B_0 \rangle$. The energy density and magnetic field decrease with redshift as larger volumes are filled. At $z = 0$, a magnetic field of $\sim 10^{-9}$ G permeates the filled overdense volumes, consistent with the results of Ryu *et al.* (2008). At a given redshift, the energy density and magnetic field are larger for smaller opening angles of the anisotropic outflows.

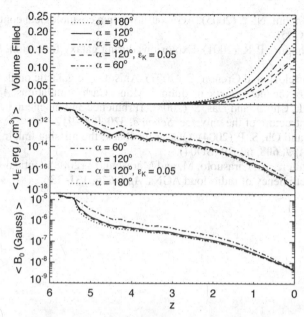

Figure 13.3 Volume fraction filled by AGN outflows (*top*), the volume weighted average of the total energy density inside outflow volumes $\langle u_E \rangle$ (*middle*), and the equipartition magnetic field within the filled volumes $\langle B_0 \rangle$ (*bottom*).

We conclude that, using our N-body simulations, the cosmological population of AGN outflows pervade 13–25% of the volume of the universe by the present. A magnetic field of $\sim 10^{-9}$ G is infused in the filled volumes at $z = 0$.

References

Barai, P. (2008). Large-scale impact of the cosmological population of expanding radio galaxies. *ApJ*, **682**, L17–L20.

Begelman, M. C., Blandford, R. D. and Rees, M. J. (1984). Theory of extragalactic radio sources. *RvMP*, **56**, 255–351.

Blustin, A. J. and Fabian, A. C. (2009). Radio constraints on the volume filling factors of AGN winds. *MNRAS*, **396**, 1732–1736.

Castor, J., McCray, R. and Weaver, R. (1975). Interstellar bubbles. *ApJ*, **200**, L107–L110.

Crenshaw, D. M., Kraemer, S. B. and George, I. M. (2003). Mass loss from the nuclei of active galaxies. *ARA&A*, **41**, 117–167.

Everett, J. E. (2007). Outflows from AGNs: a brief overview of observations and models. *Ap&SS*, **311**, 269–273.

Furlanetto, S. R. and Loeb, A. (2001). Intergalactic magnetic fields from quasar outflows. *ApJ*, **556**, 619–634 (FL01).

Ganguly, R. and Brotherton, M. S. (2008). On the fraction of quasars with outflows. *ApJ*, **672**, 102–107.

Hopkins, P. F., Richards, G. T. and Hernquist, L. (2007). An observational determination of the bolometric quasar luminosity function. *ApJ*, **654**, 731–753.

Levine, R. and Gnedin, N. Y. (2005). AGN outflows in a cosmological context. *ApJ*, **632**, 727–735 (LG05).

Martel, H. and Shapiro, P. R. (2001). Explosions during galaxy formation. *RevMexAA*, **10**, 101–108.

Pieri, M. M., Martel, H. and Grenon, C. (2007). Anisotropic galactic outflows and enrichment of the intergalactic medium. I. Monte Carlo simulations. *ApJ*, **658**, 36–51.

Ryu, D., Kang, H., Cho, J. and Das, S. (2008). Turbulence and magnetic fields in the large-scale structure of the universe. *Science*, **320**, 909–912.

Scannapieco, E. and Oh, S. P. (2004). Quasar feedback: the missing link in structure formation. *ApJ*, **608**, 62–79 (SO04).

Shankar, F., Cavaliere, A., Cirasuolo, M. and Maraschi, L. (2008). Optical-radio mapping: the kinetic efficiency of radio-loud AGNs. *ApJ*, **676**, 131–136.

14

Relativistic jets and the inhomogeneous interstellar medium

Geoffrey V. Bicknell, Jackie L. Cooper & Ralph S. Sutherland

14.1 AGN feedback from a radio galaxy perspective

The relationship between black hole mass and bulge mass (Magorrian *et al.* 1998; Gebhardt *et al.* 2000a) indicates a symbiotic relationship between the formation of supermassive black holes and galaxy formation. Silk and Rees (1998) indicated how an isotropic wind from a black hole may interact with the infalling gas in a forming galaxy to provide a natural relationship between black hole mass and bulge mass. Saxton *et al.* (2005) also showed that jets propagating through an inhomogeneous interstellar medium generate an energy-driven, more or less spherical bubble, different from the bipolar structure that is usually associated with classical radio galaxies. Thus, from our viewpoint, when we consider the interaction between jets and the interstellar medium we naturally think of gigahertz peak spectrum (GPS) and compact steep spectrum (CSS) radio galaxies as well as high redshift radio galaxies. These sources appear to be radio galaxies in the early stages of their evolution in which there is abundant evidence for strong jet–ISM interaction in the form of shock-excited emission lines and anomalous gas velocities. Given that jet power and momentum can be isotropically distributed by an inhomogeneous medium, an important issue to address is the detailed interaction between clouds and outflows in such a medium. The nature of this interaction and in particular the momentum imparted to the gas surrounding the active nucleus is going to be quite different from that envisaged by Silk and Rees (1998) and many other papers since. Therefore, in this paper we consider three main aspects of the interaction between outflows and a clumpy interstellar medium: (1) details of the interaction of jets with interstellar gas; (2) the morphology of radio galaxies at early and late times as indicators of such an interaction; (3) the details of an outflow with a single cloud.

AGN Feedback in Galaxy Formation, eds. V. Antonuccio-Delogu and J. Silk. Published by Cambridge University Press. © Cambridge University Press 2011.

165

14.2 Simulation code

The code that we have used for the simulations described here is known as *ppmlr* (piecewise parabolic method with Lagrangian remap). This code originated from the VH1 code developed by J. Blondin and colleagues (see http://wonka.physics.ncsu.edu/pub/VH-1/). The original VH1 code has been enhanced by the inclusion of thermal cooling, which is important in the interaction between dense gas and high-powered outflows. It has also been enhanced by the inclusion of code to correct numerical shock instabilities (Sutherland *et al.* 2003).

14.3 Isotropisation of jet momentum

We have already referred to the two-dimensional simulations of Saxton *et al.* (2005), which showed the isotropising effect of jet–cloud interactions. Figure 14.1 shows a snapshot from the mid-plane of a *three*-dimensional simulation, which also shows the isotropising effect of a random distribution of clouds on the passage of the jet. An energy-driven bubble is produced and this processes a much larger solid angle than would be produced by a jet in a classical radio galaxy. The clouds in this simulation are established by first generating an inhomogeneous medium with log-normal single point statistics for the density and a Kolmogorov distribution in Fourier space. The clouds were defined to have a pressure equal to that of the hot atmosphere and an average temperature of 10^4 K; the cloud medium was truncated in low-density regions when the temperature exceeded 3×10^4 K.

14.4 Jet and disk simulations

14.4.1 Initial data

Dense gas in galaxies often settles into a disk and for this reason Sutherland and Bicknell (2007) considered the passage of a jet through a thick gaseous disk supported in the vertical direction by supersonic turbulence. In the analytical model used (see below) the azimuthal velocity of the disk is almost Keplerian. A log-normal, Kolmogorov density structure was also imposed on the dense medium, but in this case the temperature was kept constant at 10^4 K and the density was truncated at the value of the surrounding hot atmosphere. In this case we utilised a gravitational potential corresponding to combined, self-consistent isothermal distributions of dark and luminous matter with different velocity dispersions for each. The *mean* density of the disk is used to apodise the initial fractal cube defined to have unit density. The mean density is defined as follows: Let $\phi(r, z)$ be the gravitational potential and $\psi(r, z) = \phi(r, z)/\sigma_D^2$ the dimensionless potential

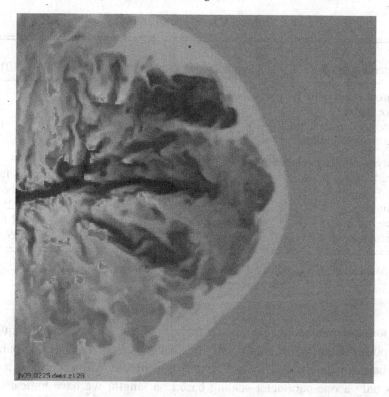

Figure 14.1 Isotropisation of a powerful jet by a clumpy medium. In this logarithmic density image the jet plasma is black and the clouds appear as the small grey regions surrounded by white (lower density) rims. The hot atmosphere is grey and the shock surrounding the expanding bubble is light grey. Parameters of this simulation: Mach number: 10; Jet/ISM density ratio: 2×10^{-3}; jet kinetic power: 1.4×10^{46} erg s^{-1}; number density of hot atmosphere: 0.1 cm^{-3}; temperature of hot medium: 4.8×10^7 K; average density of clouds: 480 cm^{-3}; temperature of clouds: $< 3 \times 10^4$ K. The resolution of this simulation is 256^3.

scaled by the dark matter velocity dispersion σ_D. Let σ_t be the turbulent velocity dispersion of the gas and T the temperature. The total velocity dispersion is σ_G where $\sigma^2 = kT/\mu m + \sigma_t^2$. Let e be the constant ratio of azimuthal to Keplerian velocity. The mean density is then given by

$$\frac{\bar{\rho}(r, z)}{\bar{\rho}(0, 0)} = \exp\left[-\frac{\sigma_D^2}{\sigma_g^2} \left[\psi(r, z) - e^2 \psi(r, 0) - (1 - e^2)\psi(0, 0) \right] \right]. \quad (14.1)$$

Further details of the description of the initial data are given in Sutherland and Bicknell (2007).

The parameters of the simulation are summarised in Table 14.1 (for the jet) and Table 14.2 for the interstellar medium. We used a non-relativistic model for the

Table 14.1 *Jet parameters*

Parameter & units	Symbol	1.0 kpc	Scaled to $x_0 =$ 0.2 kpc	5.0 kpc
Equivalent relativistic jet parameters				
†Lorentz factor	Γ	5	5	5
†Rest energy density/Enthalpy	χ	10	10	10
Velocity/Speed of light	β	0.9798	0.9798	0.9798
Hydrodynamic jet parameters				
Pressure/External pressure	ξ	1.0	1.0	1.0
Density/External density	η	2.0×10^{-3}	2.0×10^{-3}	2.0×10^{-3}
Mach number	M	25.9	25.9	25.9
†Diameter (pc)	D_{jet}	40	8.0	200.0
Kinetic luminosity	L_{jet}	2.77×10^{43}	5.54×10^{42}	1.385×10^{44}

Assigned parameters are indicated with a † symbol; others are derived.

simulations but matched the parameters to those of a relativistic jet with equivalent kinetic power (Komissarov *et al.* 1996). In adiabatic calculations a three-parameter scaling based on length, density and velocity is available. When cooling is included there is only a one-parameter scaling based on length; we have indicated three different scalings of the computational results in the tables.

14.4.2 Jet–disk simulation

At a resolution of 512 per coordinate dimension, corresponding to approximately 2 parsecs per cell for an overall scale of 1 kpc, this is one of the highest resolution simulations we have undertaken. As illustrated in Figure 14.2, the simulation reveals three main phases in the evolution of the jet–ISM interaction: (1) a flood and channel phase where the jet is severely disrupted by the clouds along its trajectory; (2) a jet breakout phase wherein the jet starts to define a fixed direction; and (3) a classical radio galaxy phase in which the radio lobe starts to resemble a normal double-lobed radio galaxy. Note, however, that in the third phase some memory remains of the earlier phases in the remnant of the spherical energy-driven bubble, which at this stage is not expanding as rapidly as in the earlier phases. We also note that, with this particular simulation, only about 4–6% of the disk gas is transported to large scales. If this type of interaction (i.e. jet interacting with porous disk) is to be important in preventing inflow into the centre of the galaxy, then some other combination of gas density and jet power may be relevant.

Table 14.2 *Halo and disk–ISM parameters*

Parameter & units	Symbol	Scaled to $x_0 =$ 1.0 kpc	0.2 kpc	5.0 kpc
Hot atmosphere				
†Virial/Gas temperature	β_h	1.0	1.0	1.0
Gas temperature (K)	T_h	1.20×10^7	1.20×10^7	1.20×10^7
Central values				
†Pressure/k (cm^{-3} K)	$p_{h,c}/k$	1.00×10^6	5.00×10^6	2.00×10^5
Pressure (dynes cm^{-2})	$p_{h,c}$	1.38×10^{-10}	6.90×10^{-10}	2.76×10^{-11}
Number Density (cm^{-3})	$n_{h,c}$	8.35×10^{-2}	4.17×10^{-1}	1.67×10^{-2}
Mass density (g cm^{-3})	$\rho_{h,c}$	8.64×10^{-26}	4.32×10^{-25}	1.73×10^{-26}
Warm disk–ISM				
Virial/Gas temperature	β_w	1200.0	1200.0	1200.0
†Gas temperature(K)	T_w	1.0×10^4	1.0×10^4	1.0×10^4
†Turbulent dispersion (km s^{-1})	σ_t	40.0	40.0	40.0
†Rotational support	E_R	0.93	0.93	0.93
Internal non–uniformity				
†Log-normal mean	μ	1.0	1.0	1.0
†Log-normal variance	σ^2	5.0	5.0	5.0
†Density power law	β	5/3	5/3	5/3
Volume of warm gas (pc^3)	V_w	2.55×10^7	2.04×10^5	3.19×10^9
Mass of warm gas ($M_\odot$)	M_w	4.67×10^5	1.87×10^4	1.17×10^7
Relative disk mass		1.0	0.04	25.0
Central values				
Pressure/k (cm^{-3} K)	$p_{w,c}/k$	1.00×10^6	5.00×10^6	2.00×10^5
†Number density (cm^{-3})	$n_{w,0}$	10.0	50.0	2.0
Mass density (g cm^{-3})	$\rho_{w,0}$	1.04×10^{-23}	5.18×10^{-23}	2.07×10^{-24}

Assigned parameters are indicated with a † symbol; others are derived.

14.5 Application to 4C31.04

The radio source 4C31.04, is classified as a compact symmetric object. This class has a lot in common with the GPS and CSS sources mentioned earlier. Indeed, the spectrum turns over at ~ 300 MHz (Kuehr *et al.* 1981) at a similar frequency to that of many CSS sources. 4C31.04 has been well studied since it is one of the closer objects of this class. Papers by Cotton, Conway, Perlman, Giroletti and colleagues (see Cotton *et al.* 1995; Conway 1996; Perlman *et al.* 2001; Giroletti *et al.* 2003; García-Burillo *et al.* 2007 as well as references therein) have established the existence of a molecular disk on pc to kpc scales, evidence for a merger in

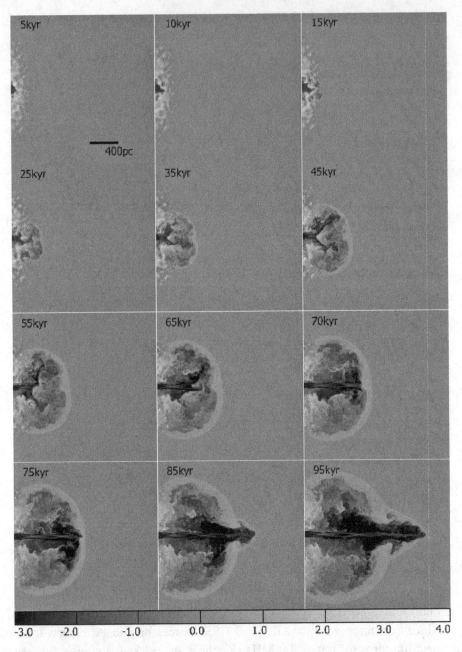

Figure 14.2 Mid-plane log-density slices of the jet–disk simulation. The panels represent the logarithm of the density in a $k = 255$ plane of the jet–disk 512^3 simulation at the various phases of the evolution. The greyscale bar shows the range of the density. The panels from 5 to 25 kyr are typical of the flood and channel phase; the panels from 35 to 55 kyr are representative of the energy-driven bubble phase; the jet-breakout phase extends from approximately 55 to 70 kyr and the classical phase extends from 75 to 95 kyr and beyond.

the relatively recent past ($\sim 10^8$ yr) and that 4C31.04 was a radio galaxy within the first few thousand years of its lifetime. There are interesting morphological similarities between the simulations described in the previous section and the lobes of this source, albeit at slightly different dynamical times. In the light of the simulations, 4C31.04 seems to be a young radio source at around the stage of jet breakout. This interpretation resolves the puzzling difference between the inferred spectral age of 3–5000 yr and the dynamical age based on the relativistic hotspot velocities. Furthermore, modelling the lobes as spherical bubbles implies a jet kinetic power $\sim (1.5-4.4) \times 10^{43}$ erg s^{-1} and an ISM central density $\sim (0.1-0.6)$ cm^{-3}. The estimated jet power is reasonable given the luminosity of the source $\sim 1.9 \times 10^{25}$ W Hz^{-1}; the estimated central density is also reasonable for the atmosphere of a giant elliptical galaxy. Fortuitously, the estimated jet kinetic power is similar to the value employed in the simulation. If AGN feedback mediated by jets is important, then it is likely that the physical processes in CSO/GPS/CSS sources such as 4C31.04 are typical of those occurring in the interaction phase. On the other hand, it should be noted that 4C31.04 is in a radio rejuvenation phase, possibly having been refuelled by a recent merger. Earlier jet–ISM events may have been more energetic and typical of the phenomena we see in high redshift radio galaxies.

14.6 Interaction of outflows with individual clouds

Given that the interaction of jets with the interstellar medium involves an outflow interacting with individual clouds, then it is interesting to examine in detail the nature of this interaction. This is based on the study by Cooper *et al.* (2009) on the interaction of starburst winds with interstellar clouds, but is relevant to any outflow–cloud interaction.

Two series of simulations are presented in this section. In both cases we have a wind of velocity 1200 km s^{-1}, number density $n_{\mathrm{w}} = 0.1$ cm^{-3} and temperature $T_{\mathrm{w}} = 5 \times 10^6$ K interacting with both spherical and fractal clouds with mean number densities $\bar{n}_{\mathrm{cl}} = 91$ cm^{-3} and $\bar{n}_{\mathrm{cl}} = 63$ cm^{-3} respectively. The spherical clouds have a radius of 5 pc, similar to the dense core of the fractal cloud. See Cooper *et al.* (2009) for further details of the initial configurations.

One interesting feature of a wind–cloud interaction is the difference between adiabatic and radiative flows reflected in the evolving morphology of the cloud. Klein *et al.* (1994) showed that an adiabatic cloud is shredded in approximately a cloud shock crossing time and this is confirmed in the upper panel of Figure 14.3. The initially spherical adiabatic cloud is heated by the shock driven into it by the wind, and since it cannot radiate away the thermal energy it expands transversely, exposing more of the cloud surface to the shredding effect of the Kelvin–Helmholtz instability. The radiative cloud, on the other hand, does not expand transversely since it radiates away the thermal energy generated by the shock. Instead, the

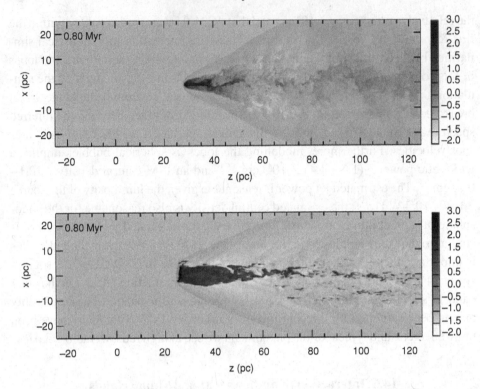

Figure 14.3 Log-density slices showing the comparison of the effect of a super-sonic wind on spherical clouds – adiabatic (upper panel) and radiative (lower panel). The greyscale bar indicates the density.

radiative cloud develops a long tail resembling the filaments observing in many starburst galaxies. One expects to see similar features in AGN driven outflows. The effect on an individual fractal cloud (extracted from one of our fractal cubes used in the jet–ISM simulations) is similar to but different from the spherical cloud simulation. The denser regions in the fractal cloud behave like individual spherical blobs, giving rise to the superposition of filamentary structures shown in Figure 14.4. There is also a distribution of velocity associated with the wind–cloud interaction and the distribution associated with the fractal cloud simulation is shown in Figure 14.5. This is the type of velocity distribution, albeit with a different scaling, that one expects to see in AGN outflow–cloud interactions. Note, however, that the velocities are substantially supersonic with respect to the internal sound speed of the clouds.

14.7 Main points

1. The radio galaxies known variously as GPS/CSS/CSO sources exhibit many of the physical processes that would have taken place in massive galaxies at earlier epochs if

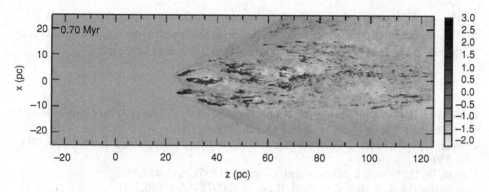

Figure 14.4 The result of a supersonic wind–fractal cloud interaction.

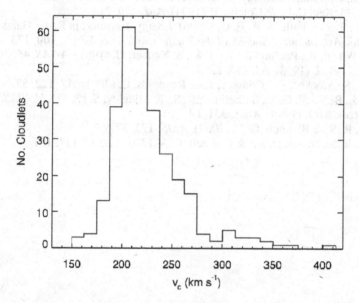

Figure 14.5 The velocity distribution produced by the wind–cloud interaction in Figure 14.4.

jet-mediated AGN feedback is important. We have illustrated this point by comparison of our simulations with a well-studied CSO 4C31.04. In Sutherland and Bicknell (2007) we have also noted features of the simulation (on a much larger scale) evident in Cygnus A.

2. The presence of a dense inhomogeneous interstellar medium forces the jet in several different directions during the 'flood and channel' phase, causing the expanding radio source to process 4π steradians of the surrounding volume.

3. The outward mass flux of dense gas is affected by the clumpiness of the medium. In the jet–ISM simulation that we have presented only ~5% of the cloud mass is driven to large radii. An exploration of jet power–cloud density–filling factor space is required in order to determine the circumstances under which jets can substantially influence infalling matter in a forming galaxy.

4. The acceleration of radiative clouds to supersonic velocities is possible.
5. The interaction of adiabatic (i.e. slowly cooling) clouds with a supersonic wind disperses the clouds. Nevertheless, the cloud material is still driven to larger radii.

References

Conway, J. E. (1996). Extragalactic Radio Sources, in *Proceedings of IAU Symposium*, eds. R. D. Ekers, C. Fanti, & L. Padrielli, **175**, 92

Cooper, J. L., Bicknell, G. V., Sutherland, R. S., & Bland-Hawthorn, J. (2009). *ApJ*, **703**, 330

Cotton, W. D., Feretti, L., Giovannini, G., *et al.* (1995). *ApJ*, **452**, 605

García-Burillo, S., Combes, F., Neri, R., *et al.* (2007). *A&A*, **468**, L71

Gebhardt, K., *et al.* (2000). *ApJL*, **539**, L13

Giroletti, M., Giovannini, G., Taylor, G. B., *et al.* (2003). *A&A*, **399**, 889

Klein, R. I., McKee, C. F., & Colella, P. (1994). *ApJ*, **420**, 213

Komissarov, S. S. & Falle, S. A. E. G. (1996). Energy Transport in Radio Galaxies and Quasars, *Astronomical Society of the Pacific Conference Series*, **100**, 173

Kuehr, H., Witzel, A., Pauliny-Toth, I. I. K., & Nauber, U. (1981). *A&AS*, **45**, 367

Magorrian, J., *et al.* (1998). *AJ*, **115**, 2285

Perlman, E. S., Stocke, J. T., Conway, J., & Reynolds, C. (2001). *AJ*, **122**, 536

Saxton, C. J., Bicknell, G. V., Sutherland, R. S., & Midgley, S. (2005). *MNRAS*, **359**, 781

Silk, J. & Rees, M. J. (1998). *A&A*, **331**, L1

Sutherland, R. S. & Bicknell, G. V. (2007). *ApJS*, **173**, 37

Sutherland, R. S., Bisset, D. K., & Bicknell, G. V. (2003). *ApJS*, **147**, 187

15

AGN feedback effect on intracluster medium properties from galaxy cluster hydrodynamical simulations

D. Fabjan, S. Borgani, L. Tornatore, A. Saro & K. Dolag

15.1 Introduction

High-quality data from X-ray satellites have established a number of facts concerning the statistical thermo- and chemo-dynamical properties of the intracluster medium (ICM) in galaxy clusters. In particular, core regions of relaxed clusters show little evidence of gas cooler than a third of virial temperature (e.g. Peterson *et al.* 2001), temperature profiles display negative gradients outside the core region (e.g. Vikhlinin *et al.* 2005; Leccardi and Molendi 2008b) and the distribution of iron in the ICM shows a negative gradient, which is more pronounced for relaxed 'cool core' clusters (e.g. Vikhlinin *et al.* 2005; Leccardi and Molendi 2008a).

The above observational properties of the ICM arise from an interplay between the cosmological scenario of building up the large-scale structure and a number of astrophysical processes (e.g. star formation, energy and chemical feedback from supernovae and AGN) taking place on much smaller scales. Such issues can be addressed using cosmological hydrodynamical simulations where the complexity of relevant astrophysical processes can be described as the result of hierarchical assembly of cosmic structures (e.g. Borgani *et al.* 2008).

The generally accepted solution to the shortcomings of simulations is represented by AGN feedback. The presence of cavities in the ICM at the cluster centre is considered as the fingerprint of the conversion of the mechanical energy associated with AGN jets into thermal energy (and possibly in a non-thermal content of relativistic particles) through shocks (e.g. McNamara and Nulsen 2007).

The aim of this work was to perform a detailed analysis of cosmological hydrodynamic simulations of galaxy clusters, which have been carried out with the GADGET-2 code (Springel 2005), by combining the AGN feedback mode

AGN Feedback in Galaxy Formation, eds. V. Antonuccio-Delogu and J. Silk. Published by Cambridge University Press. © Cambridge University Press 2011.

described by Springel *et al.* (2005) with the SPH implementation of chemo-dynamics presented by Tornatore *et al.* (2007).

The results presented here concern the thermodynamical and chemical properties of the hot diffuse intracluster gas and the effect that SN and AGN feedback have on the chemical enrichment of the ICM.

15.2 The simulations

Our set is composed of 16 simulated clusters (Dolag *et al.* 2008). To obtain them we extracted nine Lagrangian regions from a dark matter (DM) only simulation with a box size of $479h^{-1}$ Mpc (Yoshida *et al.* 2001), performed for a flat Λ cold dark matter (ΛCDM) cosmological model ($\Omega_m = 0.3$, $h_{100} = 0.7$, $\sigma_8 = 0.9$). The extracted regions contain in their centres five isolated galaxy groups ($M_{200} \simeq 1 \times 10^{14}h^{-1}M_\odot$) and four massive clusters ($M_{200} \geq 1$–$2.2h^{-1}M_\odot$), beside eight satellite clusters ($M_{200} \geq 5 \times 10^{13}h^{-1}M_\odot$). Each region was resimulated using the zoomed initial condition (ZIC) technique (Tormen *et al.* 1997). The massive and satellite clusters were simulated at the basic resolution with $m_{DM} \simeq 1.9 \times 10^8$ $h^{-1}M_\odot$ and $m_{gas} = 2.8 \times 10^7$ $h^{-1}M_\odot$, while the small groups were simulated at $6\times$ higher mass resolution.

Simulations were preformed using the TreePM-SPH GADGET 2 code (Springel 2005). All the simulations include a metallicity-dependent radiative cooling (Sutherland and Dopita 1993), heating from a uniform time-dependent ultraviolet background (Haardt and Madau 1996) and the effective model by Springel and Hernquist (2003) for the description of star formation. Our simulations also include the detailed description of stellar evolution and chemical enrichment by Tornatore *et al.* (2007) (where more details can be found). Metals are produced by SN-II, SN-Ia and intermediate and low-mass stars in the asymptotic giant branch (AGB hereafter). More details can be found in Fabjan *et al.* (2009).

15.2.1 Feedback models

For the whole cluster set we performed four series of runs, corresponding to as many prescriptions for energy feedback:

(a) no feedback (NF),
(b) galactic winds (W),
(c) standard implementation of AGN feedback (AGN-1),
(d) modified version of the AGN feedback (AGN-2).

In the first case (a) neither galactic winds nor AGN feedback is included.

Galactic winds (b) are included by following the model of Springel and Hernquist (2003). In this case energy released by SN-II triggers galactic winds, with mass upload rate assumed to be proportional to the star formation rate, $\dot{M}_W = \eta \dot{M}_\star$. We set the wind velocity to $v_w = 500 \, \text{km s}^{-1}$ and $\eta = 2$ for the wind mass upload.

For AGN-1 (c) we include in our simulations the effect of feedback energy released by gas accretion onto supermassive black holes (BHs), following the scheme originally introduced by Springel *et al.* (2005), to which we refer for a more detailed description. In this model BHs are represented by collisionless sink particles that grow via gas accretion and through mergers with other BHs during close encounters. In every new dark matter halo above a certain mass threshold a BH of mass $10^5 h^{-1} M_\odot$ is deposited, provided the halo does not contain any BHs yet. Once seeded, each BH can then grow by local gas accretion, with a rate given by $\dot{M}_{BH} = \min\left(\dot{M}_B, \dot{M}_{Edd}\right)$, where $\dot{M}_B$ is the accretion rate estimated with the Bondi–Hoyle–Lyttleton formula (Hoyle 1939; Bondi 1944, 1952), while $\dot{M}_{Edd}$ is the Eddington rate. The latter is inversely proportional to the radiative efficiency ϵ_r, which gives the radiated energy in units of the energy associated with the accreted mass. A fraction ϵ_f of the radiated energy is thermally coupled to the surrounding gas, so that $\dot{E}_{feed} = \epsilon_r \epsilon_f \dot{M}_{BH} c^2$ is the rate of energy feedback provided. Following Springel *et al.* (2005), we use $\epsilon_r = 0.1$ and $\epsilon_f = 0.05$ as a reference value.

Following Sijacki *et al.* (2007), we assume for AGN-2 (d) that a transition from a 'quasar' phase to 'radio' mode takes place whenever the accretion rate falls below $\dot{M}_{BH}/\dot{M}_{Edd} < 10^{-2}$, with an increase of the feedback efficiency from $\epsilon_f = 0.05$ to $\epsilon_f = 0.2$. To provide a more uniform distribution of energy we chose to distribute energy around the BH with a top-hat kernel, mimicking the effect of inflating bubbles in correspondence with the termination of the AGN jets.

15.3 Temperature profiles

A number of comparisons between observed and simulated temperature profiles of galaxy clusters have clearly established that simulations are indeed remarkably successful at reproducing data at relatively large radii, $R \gtrsim 0.2 R_{180}$, where the effect of cooling should be relatively unimportant. The same simulations have a much harder time predicting realistic profiles within cool-core regions (see Borgani *et al.* 2008 for a recent review). In this regime, radiative simulations systematically produce steep negative temperature profiles, at variance with observations, as a consequence of the lack of pressure support caused by overcooling. While feedback associated with SNe has been proved not to be successful, AGN feedback is generally considered as a likely solution for simulations to produce realistic cool cores.

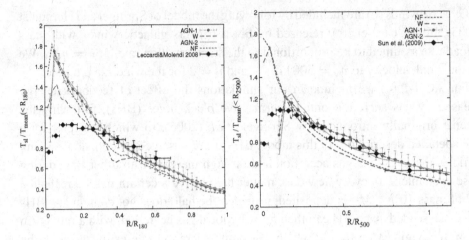

Figure 15.1 Temperature profiles of galaxy clusters (*left*) and groups (*right*).

We present in Figure 15.1 the comparison between simulated and observed temperature profiles for galaxy clusters with $T \gtrsim 3\,\mathrm{keV}$ (left panel) and for poorer clusters and groups with $T \lesssim 3\,\mathrm{keV}$ (right panel). Observational results are taken from Leccardi and Molendi (2008b) and Sun *et al.* (2009), respectively. As for rich clusters, none of the implemented feedback schemes is capable of preventing the temperature spike at small radii, while all the models provide a temperature profile quite similar to the observed one at $R \gtrsim 0.3 R_{180}$. For groups, instead, both schemes of AGN feedback provide results that go in the right direction. While galactic winds are not able to regulate the steep negative temperature gradients, the AGN-1 and the AGN-2 feedback schemes pressurise the ICM in the central regions, thus preventing adiabatic compression in inflowing gas. This result confirms that a feedback scheme not related to star formation goes indeed in the right direction of regulating the thermal properties of the ICM in cool-core regions. The AGN feedback schemes implemented in our simulations do an excellent job at the scale of galaxy groups, while they are not efficient enough at the scale of massive clusters.

15.4 Metal enrichment of the ICM

The study, through X-ray spectroscopy, of the content and distribution of metals in the intracluster plasma provides invaluable hints on the connection between the process of star formation, which occurs on small scales within galaxies, and the processes that determine the thermal properties of the ICM. The former affects the quantity of metals that are produced by different stellar populations, while the latter gives us insights on gas-dynamical processes, related both to the gravitational

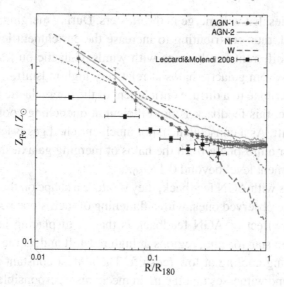

Figure 15.2 Z_{Fe} profiles of massive galaxy clusters.

assembly of clusters and to the feedback mechanisms that displace metal-enriched gas from star-forming regions.

Profiles of iron abundance are the standard way to characterise the pattern of ICM metal enrichment from observations. We show in Figure 15.2 the emission-weighted iron abundance profiles for the four adopted feedback schemes. Profiles are obtained by averaging over the four simulated clusters with $T_{sl} > 3$ keV, compared with observational results. For reasons of clarity we report the computed 1σ scatter only for the AGN-1 runs. Simulation predictions are compared with the observational results by (Leccardi and Molendi 2008a). All the abundance values are scaled to the solar abundances provided by Grevesse and Sauval (1998).

All our simulations predict the presence of abundance gradients in the central regions at least for $R \lesssim 0.1R_{180}$. The lowest enrichment level is actually found for the NF run, where the highly efficient cooling selectively removes the most enriched gas, which has the shortest cooling time, thus leaving metal-poorer gas in the diffuse phase. Runs with galactic winds (W) and AGN feedback (AGN-1 and AGN-2) are instead able to better regulate gas cooling in the central region, thus allowing more metal-rich gas to survive in the hot phase. For this reason, W and AGN runs predict profiles of Z_{Fe} that are steeper than for the NF runs, in the central regions, $\lesssim 0.1R_{180}$.

The different nature of SN-powered winds and AGN feedback leaves a much more clear imprint at larger radii. As for the NF runs, they produce a rather high level of enrichment out to $\sim 0.3R_{180}$, while rapidly declining at larger radii. In this model, the high level of star formation provides a strong enrichment of the gas in the halo of galaxies that will merge in the clusters. During merging, this gas is ram-pressure stripped, thus contributing to increase the enrichment level of the ICM. The situation is different for the runs with winds. Galactic outflows are efficient in displacing gas from galactic halos at relatively high redshift, $z \gtrsim 2$, when they significantly contribute to a diffuse enrichment of the intergalactic medium (IGM). At the same time, this feedback is not efficient at quenching cooling of enriched gas at low redshift. As a consequence, not much enriched gas is left to be stripped by the hot cluster atmosphere from the halos of merging galaxies, thus explaining the lower enrichment level beyond $0.1 R_{180}$.

As for the runs with AGN feedback, they produce a shape on the abundance pro-files closest to the observed ones, with a flattening of their slope beyond $\sim 0.2 R_{180}$. In this case, the effect of AGN feedback is that of displacing large amounts of enriched gas from star-forming regions at high redshift and, at the same time, effi-ciently suppressing cooling at low redshift. The almost constant level of Z_{Fe} out to R_{180} and beyond witnesses that the main mechanism responsible for enrichment in this case is not ram-pressure stripping, whose efficiency should decline with cluster-centric radius, but the diffuse accretion of pre-enriched IGM. We note that the slightly higher Z_{Fe} in the AGN-2 scheme with respect to the AGN-1 is due to the effect of the more efficient radio-mode feedback, which helps in displacing enriched gas from massive galaxies.

Although the models with AGN feedback produce the correct shape of the iron abundance profiles, their normalisation is generally higher than for the observed ones. This overproduction of iron could be due to the uncertain knowledge of a number of ingredients entering the chemical evolution model implemented in the simulation code (e.g. stellar yields, see Tornatore et al. (2007)). We believe that the shape of the abundance profiles, instead of their amplitude, should be considered as the relevant observational information to be used to study the impact that different feedback mechanisms have on the ICM enrichment pattern.

15.5 Conclusions

We have presented the analysis of an extended set of cosmological hydrodynamical simulations of galaxy clusters aimed at studying the different effects that stellar and AGN feedback have on the thermal and chemo-dynamical properties of the intra cluster medium (ICM). For this purpose we used a version of the Tree-SPH GADGET-2 code (Springel 2005), with a detailed description of chemical evolution (Tornatore et al. 2007). All cluster simulations of this set have been run using different prescriptions for the feedback: without including any efficient feedback (NF runs); including only the effect of galactic winds powered by supernova (SN) feedback (W runs); and including two different prescriptions of AGN feedback

(AGN-1 and AGN-2) based on modelling gas accretion supermassive black holes (BHs) hosted within resolved galaxy halos (Springel *et al.* 2005).

The main results of our analysis can be summarised as follows.

(a) AGN feedback is quite efficient in pressurising gas in the central regions of galaxy groups, thereby generating temperature profiles that are in reasonable agreement with the observed ones. However, temperature profiles in the core regions of massive clusters are still too steep, even after including AGN feedback.

(c) The presence of AGN feedback generates a rather uniform and widespread pattern of metal enrichment in the outskirts of clusters. This is the consequence of the improved efficiency, with respect to the runs without AGN feedback, at extracting at high redshift highly enriched gas from star-forming regions and, therefore, enhancing metal circulation in the intergalactic medium.

(f) Radial profiles of iron abundance are predicted to be too steep at $R \gtrsim 0.1 R_{180}$ in runs including stellar feedback. Their shape is in much better agreement with the observed ones when including AGN feedback.

Our analysis lends further support to the idea that a feedback source associated with gas accretion onto supermassive BHs is required by the observational properties of the ICM (McNamara and Nulsen 2007).

The presented results further demonstrate that different astrophysical feedback sources leave distinct signatures on the pattern of chemical enrichment of the ICM. These differences are much more evident in the outskirts of galaxy clusters, which retain memory of the past efficiency that feedback models had in displacing enriched gas from star-forming regions and in regulating star formation itself. However, characterisation of thermal and chemical properties in cluster external regions requires X-ray telescopes with large collecting areas and excellent control of the background. A detailed knowledge of the ICM out to the cluster virial boundaries has to await the advent of the next generation of X-ray telescopes (Giacconi 2009).

Acknowledgements

We would like to thank Volker Springel for having provided us with the non-public version of GADGET-2. We thank Alberto Leccardi and Ming Sun for having kindly provided tables of observational data points. We acknowledge useful discussions with Francesca Matteucci, Pasquale Mazzotta, Silvano Molendi, Ewald Puchwein, Elena Rasia, Debora Sijacki, Paolo Tozzi and Matteo Viel. Simulations have been carried out at the CINECA (Bologna, Italy), with CPU time assigned thanks to an INAF-CINECA grant and the agreement between CINECA and the University of Trieste, and at the Computing Centre at the University of Trieste. The work has been partially supported by the INFN PD-51 grant, INAF-PRIN06 grant, ASI-AAE Theory grant and by PRIN-MIUR 2007 grant. KD acknowledges the financial support of the HPC-Europa Transnational Access program and the hospitality of CINECA.

References

Bondi, H., & Hoyle, F. (1944). *MNRAS*, **104**, 273

Bondi, H. (1952). *MNRAS*, **112**, 195

Borgani, S., Diaferio, A., Dolag, K., & Schindler, S. (2008). *Space Science Reviews*, **134**, 269

Dolag, K., Borgani, S., Murante, G., & Springel, V. 2008, arXiv:0808.3401

Fabjan, D., Borgani, S., Tornatore, L., *et al.* (2010). *MNRAS*, **401**, 1670

Giacconi, R. (2009). *Astronomy*, **2010**, 90

Grevesse, N., & Sauval, A. J. (1998). *Space Science Reviews*, **85**, 161

Haardt, F., & Madau, P. (1996). *ApJ*, **461**, 20

Hoyle, F., & Lyttleton, R. A. (1939). *Proceedings of the Cambridge Philosophical Society*, **35**, 405

Leccardi, A., & Molendi, S. (2008a). *A&A*, **487**, 461

Leccardi, A., & Molendi, S. (2008b). *A&A*, **486**, 359

McNamara, B. R., & Nulsen, P. E. J. (2007). *ARA&A*, **45**, 117

Peterson, J. R., *et al.* (2001). *A&A*, **365**, L104

Sijacki, D., Springel, V., di Matteo, T., & Hernquist, L. (2007). *MNRAS*, **380**, 877

Springel, V., Di Matteo, T., & Hernquist, L. (2005). *MNRAS*, **361**, 776

Springel, V. (2005). *MNRAS*, **364**, 1105

Springel, V., & Hernquist, L. (2003). *MNRAS*, **339**, 289

Sun, M., Voit, G. M., Donahue, M., *et al.* (2009). *ApJ*, **693**, 1142

Sutherland, R. S., & Dopita, M. A. (1993). *ApJS*, **88**, 253

Tormen, G., Bouchet, F. R., & White, S. D. M. (1997). *MNRAS*, **286**, 865

Tornatore, L., Borgani, S., Dolag, K., & Matteucci, F. (2007). *MNRAS*, **382**, 1050

Vikhlinin, A., Markevitch, M., Murray, S. S., *et al.* (2005). *ApJ*, **628**, 655

Yoshida, N., Sheth, R. K., & Diaferio, A. (2001). *MNRAS*, **328**, 669

16

Physics and fate of jet-related emission line regions

Martin G. H. Krause & Volker Gaibler

16.1 Introduction

At redshifts above $z \gtrsim 0.5$ extragalactic jet sources are commonly associated with extended emission line regions (for a review see McCarthy 1993; Miley and De Breuck 2008). The most prominent emission line is the hydrogen Lyman α line, but other typical nebular emission lines have also been found. These regions are up to 100 kpc in extent, anisotropic and preferentially aligned with the radio jets (*alignment effect*). Their properties correlate with those of the radio jets: smaller radio jets (<100 kpc) have more extended emission line regions with larger velocity widths (1000 km s^{-1}) that are predominantly shock ionised, as diagnosed from their emission line ratios. Larger radio jets (>100 kpc) have emission line regions even smaller than 100 kpc. Their turbulent velocities are typically about 500 km s^{-1} and the dominant excitation mechanism is photoionisation. The physical function of these emission line regions can be compared to a detector in a particle physics experiment: in both cases a beam of high-energy particles hits a target. Analysis of the interactions in the surrounding detector, or in astrophysics the emission line gas, provides information about the physical processes of interest. For the astrophysical jets, the information one would like to obtain from such analysis concerns two traditionally separate branches of astrophysics.

The considerable energy release that may be associated with the jet phenomenon is received by a large reservoir of gas surrounding the host galaxy. Compression and heating of that gas may influence the star formation history of the host galaxy, in both triggering and suppression of star formation (Best *et al.* 1996; Dey *et al.* 1997; Tortora *et al.* 2009). Also, the jets transfer a bulk momentum to the emission line gas surrounding the galaxy. The observed speeds in the emission line gas are typically of the order of the escape velocity of the host system (Nesvadba *et al.*

AGN Feedback in Galaxy Formation, eds. V. Antonuccio-Delogu and J. Silk. Published by Cambridge University Press. © Cambridge University Press 2011.

2008). This may lead to observable effects in the spectral energy distribution of that galaxy at later times (e.g. Tortora et al. 2009), and might even be part of the answer to the question why the mass of the supermassive black holes correlates with properties of the host galaxies (e.g. Häring and Rix 2004).

The other branch is the physics of active galactic nuclei. Extragalactic jets originate in the immediate vicinity of supermassive black holes. How the jet production is really happening is not entirely clear yet, partly because the horizon scale, which is the characteristic size of a black hole, cannot be resolved by any instrument in any object. Also, an adequate treatment of the Kerr metric, which permits angular momentum exchange in both ways between the black hole and the plasma in the surrounding accretion disk via magnetic torques, is still a challenge to current magnetohydrodynamic (MHD) studies of accretion physics, though progress is being made (Hawley et al. 2007). Within the theory of general relativity, black holes have, in principle, three basic properties: mass, angular momentum and charge. As usual in astrophysics, the charge is assumed to be negligible. Black hole masses are in general well measured. Because the horizon scale is proportional to the mass, the mass, together with the accretion rate, sets the scale for the overall energy output. Jet production might be coupled to the black hole's angular momentum and magnetosphere. The latter may be considerably different for different environmental conditions, especially accretion rates. Some black holes develop a strong quasar activity having at the same time powerful jets, while some do not. Common sense suggests that, since masses and accretion rates do not seem to make a difference, the black hole's angular momentum, together with the magnetosphere, might be the decisive point (e.g. Blandford and Znajek 1977; Camenzind 1991; McNamara et al. 2009). Important evidence for or against this hypothesis might in future come from correlating reliable measurements of black hole spins and jet powers. Black hole spins have been measured from relativistically deformed iron lines of near horizon material (e.g. Müller and Camenzind; Young et al. 2005 and references therein). The values found include rapidly spinning black holes close to the limit allowed by the Kerr metric (cosmic censorship conjecture). Jet powers have so far been constrained from the interaction with the ambient gas, mainly from X-ray data (e.g. Krause 2005a; Bîrzan et al. 2008), but first attempts have been made to use the energy in the emission line gas for this (Nesvadba et al. 2008).

Jet beams are nearly dissipationless, and transport various kinds of information to large scales, where they might, in principle, be deduced from careful observation and analysis of their interaction with the environment: the jet's power, mass flux, particle composition, net electric current, and also the sense of the toroidal as well as the direction of the axial magnetic field should be conserved along the jet beam. Extended emission line regions may, in principle, serve as calorimeters. Indeed, the emission line power is correlated with the radio power (e.g. McCarthy 1993). If shock ionisation is firmly deduced from the emission line ratios, we may conclude that the power source of the radiation is the jet. Also, the kinetic energy in the emission line gas, which can now be quite accurately determined from integral field spectroscopy, is a lower bound for the energy released by the jet. For two powerful radio galaxies at a redshift of about two, Nesvadba *et al.* (2008) give the observed radio power, power in the Hα line and the power required to account for the kinetic energy in the emission line gas; each value is a few times 10^{45} erg s^{-1}.

Here, we present hydrodynamics (HD) and magnetohydrodynamics simulations of the interaction of powerful radio jets with their environment, with an emphasis on the cold gas ($T < 10^6$ K), which includes the emission line gas. We constrain the locus of the emission line regions as well as the fraction of the jet power they receive by global simulations in Section 16.2. We present local box simulations of the multi-phase turbulence on small scales in Section 16.3, which can be used to infer properties of the emission line gas such as kinematics or condensation rate from the hot gas entrained into the radio cocoons. We discuss our results and present the conclusions in Section 16.4

16.2 Global jet simulations

We have recently carried out a parameter study of global jet simulations with the MHD code NIRVANA (Ziegler and Yorke 1997; Gaibler *et al.* 2008, 2009). The jets are injected at Mach 6 into a constant ambient density environment. Here we use the results for the simulation with a density ratio of 10^{-3}. We use an innovative magnetic field configuration: a helical magnetic field constrained to the jet material, which is almost zero in the ambient medium. The global morphology of this simulations is shown in Figure 16.1. The narrow jet beams propagate essentially undisturbed from the centre of the image out to the edges of the low-density cocoon. The supersonic beams are shocked there at the so-called Mach disk. The kinetic energy is transfered into thermal energy and produces the hot spots there. Because of the large density contrast, the excess pressure of the hot spots drives comparatively slow bow shocks into the surrounding gas, which may be seen in Figure 16.1 as an elliptical structure surrounding the source. Because of the slow bow shock expansion, the jet plasma has to flow backwards from the hot spots. At the same time it expands sideways till it reaches pressure equilibrium with the surrounding layer of shocked ambient gas. The resulting width of the low-density cocoon is anti-correlated to the jet/ambient density ratio. The large cocoon widths observed in extragalactic radio jets usually require light jets, often the inferred jet density is three to four orders of magnitude below the ambient density. The simulations have shown that very light jets produce a thick layer of shocked ambient gas, and weak bow shocks. Where the bow shock can be identified in X-ray data, both predictions are confirmed (e.g. Nulsen 2005).

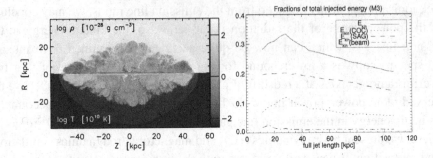

Figure 16.1 Left: MHD simulation of a very light (density ratio 10^{-3}) jet. Upper part shows the density, lower part the temperature. Right: Kinetic energy as a fraction of the total injected energy, for all the simulation and selected regions explained in the text. The remaining energy is thermalised.

The contact surface between the low-density cocoon and the shocked ambient gas layer, though partly stabilised by magnetic fields, is ragged by Kelvin–Helmholtz instabilities. Here, the ambient gas is entrained into the radio cocoon, where it is accelerated by the turbulent cocoon motions. Turbulence-enhanced cooling, as proposed below, may enable such gas to cool down and contribute to the observed emission line halos.

An important question is, where is the emission line gas actually located with respect to the structures in the jet simulation? Many authors have assumed that it should be the layer of shocked ambient gas, where the emission line gas actually resides. An alternative location would be the low-density jet cocoon (cocoon hereafter). From observations the emission line gas is usually volume filling and does not show any kind of shell structure. The latter would be expected if the emission line gas were located in the shocked ambient gas layer. Therefore, the morphology strongly points to the cocoon as the emission line locus. For an inclination of 70 degrees from the jet axis, we have calculated the position–velocity diagrams for the cocoon and shocked ambient gas layer, where the two have been separated with a tracer technique (Figure 16.2). The plot for the shocked ambient gas displays the expected velocities, as they are similar to the sound speed in the ambient gas. However, the approaching and the receding parts of the shocked ambient gas layer occupy clearly separated regions, which are usually not observed. A nice feature is the slope of the curve, which might be consistent with the observed large-scale velocity gradients. The cocoon gas occupies a coherent region. The velocities here refer to a numerical mixture of relativistic jet and hot entrained ambient plasma. Below, we will show that the multi-phase turbulence produces a characteristic density–velocity relation, which can account for the observed velocities in the emission line gas.

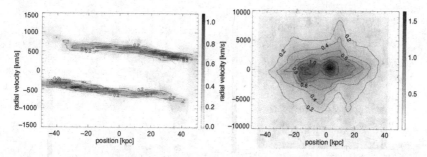

Figure 16.2 Position–velocity diagram for the simulation of Figure 16.1. Left: Cells in the shocked ambient gas. Right: Cells in the cocoon.

A remaining problem for the cocoon locus is the bulk velocities: cocoons in light jets flow backwards from the hot spots, in the observer's frame. Emission line regions are observationally inferred to move outwards. We will address this problem further below. We have calculated the kinetic energy available for cloud acceleration in the two respective regions in question (Figure 16.1, right). For our reference simulation, the kinetic energy in the cocoon is in the range 5–10 per cent of the total injected energy. For the shocked ambient gas layer, the proportion is 15–20 per cent. These numbers are anti-correlated with the jet/ambient density ratio. Within the accuracy of the measurements, both situations would require a similar conversion between the measured kinetic energy in the emission line regions and the total jet power.

16.3 Local simulations of multi-phase turbulence

The actual emission line gas component cannot be resolved well in global jet simulations. We have therefore done local 3D HD simulations, including optically thin cooling, also with NIRVANA. We essentially simulate the Kelvin–Helmholtz instability with a density ratio of 10 000 and a dense cloud (1 and 10 cm^{-3}) in the middle. The initial Mach number is 0.8 (80) in the hot (intermediate) phase, respectively. In a series of simulations, we vary the cloud density and the temperature of the intermediate (ambient) temperature phase. The cloud is soon disrupted and the non-linear evolution of the instability leads to multi-phase turbulence (Figure 16.3, left). There are several numerical effects that may play a role here. They are discussed along with more details about the simulations in Krause (2008). We believe, however, that the results we focus on here are not dominated by the numerical issues.

We show a typical temperature histogram in Figure 16.4 (left). The intermediate temperature gas is still very visible at 5×10^6 K. Numerical mixing produces the

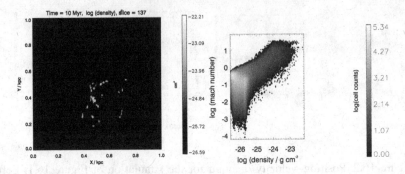

Figure 16.3 Left: Density slice through the μ1T5HR (see caption of Figure 16.5 for a description of the numbers) simulation at 10 Myr. Right: Mach number–density histogram for the same simulation.

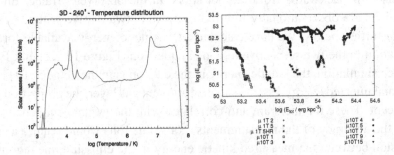

Figure 16.4 Left: Temperature histogram (μ1T5HR). Shock excitation and cooling together produce the pronounced peak at 10^4 K, here identified with the emission line gas. Right: Energy in the emission line gas versus total energy in the simulation. The simulations with low mass load have a lower fraction of their energy in the emission line gas. The time evolution is from high total energy to lower total energy.

flat regions at both sides of this peak. Towards the lower temperature end, the cutoff is given by the resolution limit of the simulation. Near 10^4 K, a prominent peak is found. This may be interpreted as shock-excited emission line gas. In the following analysis we define gas with a temperature between 10 000 K and 20 000 K as emission line gas.

Figure 16.3 (right) shows the Mach number–density histogram. The Mach number is not correlated to the density at low densities, where the cooling time is much longer than the simulation time. When cooling is important, the Mach number is proportional to the square root of the density. The result is identical to the 2D case (Krause and Alexander 2007). It implies that the velocity distribution of all gas colder than $\sim 10^6$ K should be the same. Also, from the critical density when cooling becomes important down to the density of jet material, the velocity scales

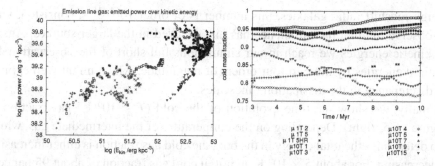

Figure 16.5 Left: Radiative power emitted by the emission line gas (temperature in the range from $10\,000$ K to $20\,000$ K) vs. kinetic energy of this gas. Different symbols are used for different runs shown at many times, as indicated in the legend. μxTy denotes an initial cloud density of $x\, m_p$ cm^{-3} and an initial temperature of the intermediate component of $y10^6$ K. Right: Cold gas ($T < 10^6$ K) fraction over time.

with the square root of the density. The velocity of the backflow of the jet plasma in the cocoon should still be of order of the speed of light. The critical density is probably not very much higher than the ambient density in these strongly star-forming galaxies. The observed width of the emission line gas then requires jet/ambient density ratios of order 10^{-4}–10^{-3}, similar to what can be deduced from X-ray observations of sources at low redshift (Krause 2003, 2005a).

We show the energy fraction in the emission line gas in Figure 16.4 (right). Due to the radiative dissipation, the time evolution for each of the different simulations is from high to lower total energy. Except for the highest ambient temperature run, the energy in the emission line gas first rises, then, for many simulations, levels off at a constant energy. For those simulations, the energy in the emission line gas is a few tenths of the total energy. This fraction depends strongly on the mass load, with all the low mass load simulations having a much lower energy fraction in the emission line gas. Our simulations have a mass load of $\sim 10^5 M_\odot$ kpc^{-1}, roughly three orders of magnitude less than observed extended emission line regions. All this suggests that the latter have most of their turbulent energy in the emission line gas. There could also be a significant fraction in a yet colder component, which our simulations do not resolve well enough.

Our simulations also suggest a relation between the kinetic energy in the emission line gas (far more than the thermal energy in this gas phase) and the radiated power (Figure 16.5, right). At low kinetic energy, about 10^{12} erg are needed to radiate at 1 erg s^{-1}, or equivalently, the radiative dissipation timescale, which is the relevant dissipation timescale for multi-phase turbulence, is 10^{12} s. At higher energy, the dissipation timescale has a wider spread for the different simulations with a tendency to increase to about 10^{13} s. For the observed emission line regions

in high redshift radio galaxies, this number is of order 10^{14-15} s, consistent with
the observation that the turbulence has just decayed in the larger sources. Again,
the kinetic energies we reach in our simulations fall short of the observationally
required energies by three to four orders of magnitude, which might well explain
the discrepancy in the dissipation timescales.

Finally, we show the time evolution of the cold ($T < 10^6$ K) gas fraction in
(Figure 16.5, right). Depending on the temperature of the intermediate gas, which
also determines the total energy in the box, the cold gas fraction is either increasing
or decreasing. For about 5×10^6 K an initial cold gas fraction of about 95 per cent
seems to be right. The result depends on the numerical resolution in the way that
we underestimate the cold gas fraction. In summary, for ambient temperatures of
up to 5×10^6 K, in equilibrium, most of the gas mass should be in the cold phase.
From these results, it seems plausible in principle that much of the emission line
gas is cooled from entrained warmer ambient gas, with the help of some initial cold
gas filaments.

16.4 Discussion and conclusions

We have presented global jet simulations, as well as local box simulations of multi-
phase turbulence. Regarding the locus of the emission line gas in high redshift
radio galaxies, from the global simulations, both the shocked ambient gas and the
radio cocoon have problems. The most severe for the shocked ambient gas is the
expected shell morphology and the split nature of the position–velocity diagram.
While dust may play some role in individual sources, the rarity of these features in
observed sources argues against this scenario. The cocoon model requires a detailed
model for the multi-phase turbulence in the cocoon. We give such a model with our
box simulations, and show that kinematics and radiated power can be explained.
We can even give a possible origin for the emission line gas; namely, turbulence-
enhanced cooling on a small seed cold gas fraction. A remaining problem is the
bulk velocities. While the simulations predict inward motion during the active jet
phase, the observations, based on the Laing–Garrington effect, show outflowing
gas. Perhaps a way to make sense of this would be to assume that much of the
emission line gas is actually not gas entrained from the halo into the cocoon
in a warm phase, but rather initially kinematically cold gas within the galaxy.
If the turbulent radio cocoon interacted with a massive gas disk, such as those
observed at similar redshifts by Förster Schreiber *et al.* (2009), one would expect
a turbulent diffusion into the radio lobes, with roughly the observed bulk flow
properties. This conclusion is supported by a consideration of the momentum
in the emission line gas and the jet. For one bubble of emission line gas the

momentum is

$$P_{elg} = 10^{51} \, g \, cm \, s^{-1} M_{elg,10} \, v_{elg,500} \,, \tag{16.1}$$

where $M_{elg,10}$ is the mass of the emission line gas in units of $10^{10} M_\odot$ and $v_{elg,500}$ is the bulk outflow velocity in units of 500 km s^{-1} (Nesvadba *et al.* 2008). If Cygnus A is regarded as a prototype for FR II radio sources, then the total mass driven through the jet beam for a ~100 kpc sized source would be of order

$$M_{jet} = 10^4 \, M_\odot \, \dot{M}_{-3} t_7 \,, \tag{16.2}$$

where $\dot{M}_{-3}$ is the jet's mass flux in units of $10^{-3} M_\odot$ yr^{-1}, and t_7 is the age of the source in units of 10 Myr. The total momentum delivered by the jets is hence typically of order

$$P_{jet} = 10^{48} \, g \, cm \, s^{-1} M_{jet,4} \Gamma_3 \,. \tag{16.3}$$

Again, the mass through the jet beam is given in convenient units of $10^4 M_\odot$ and the jet's bulk Lorentz factor is scaled to three. So, unless the jets are extremely relativistic, which seems unlikely, the jet's momentum falls short of that required to accelerate the emission line gas by orders of magnitude. So, while the jet energy is high enough to cause turbulent motion, the momentum is too small to be responsible for the directed motions. Further support for this scenario comes from the HI shells surrounding small high-redshift radio galaxies only (Krause 2005b). They have a typical diameter of 50 kpc, and a typical velocity of 200–300 km s^{-1}. The energy required to drive these shells can therefore be calculated precisely and is of order one supernova per year, corresponding to star formation rates in excess of 100 $M_\odot$ yr^{-1}. This is actually found in observations (Förster Schreiber *et al.* 2009). If this is indeed evidence for the end of a starburst, one could interpret the emission line halos as the missing observational evidence for the late phase in galaxy merger simulations with active black holes (Di Matteo *et al.* 2005). Due to the shape of the cavities drilled by the jets, the expelled gas is not distributed isotropically, but aligned with the radio jets.

What happens to the gas when the active jet phase terminates? We have carried out simulations of the long-term effect of radio sources, representing the emission line gas by tracer particles (Heath *et al.* 2007). Powerful jets set up a long-term gas convection that may persist for many Gyr. From these simulations, we predict the emission line gas to be distributed in the ambient gas, contributing to the halo metallicity.

Focusing again on the black hole, the emission line regions give power estimates close to the emitted radio power. Together they approach 10^{46} erg s^{-1}. The jets must have considerably more power than this to keep propagating. We showed above that overall of order 1 per cent, maybe up to 10 per cent of the total jet energy should be expected to appear as kinetic energy in the emission line gas; hence, extended emission line regions point to kinetic jet powers of order 10^{47} erg s^{-1} or greater. This is of the order of the Eddington power for a $10^9 M_\odot$ black hole. If the accretion state of the central quasar could be sufficiently constrained, one might be able to infer whether the jet is powered by accretion directly or by the spinning down of the central black hole.

References

Best, P. N., Longair, M. S., & Röttgering, H. J. A. (1996). Evolution of the aligned structures in $z \sim 1$ radio galaxies. *MNRAS*, **280**, L9

Bîrzan, L., McNamara, B. R., Nulsen, P. E. J., Carilli, C. L., & Wise, M. W. (2008). Radiative efficiency and content of extragalactic radio sources: toward a universal scaling relation between jet power and radio power. *ApJ*, **686**, 859

Blandford, R. D., & Znajek, R. L. (1977). Electromagnetic extraction of energy from Kerr black holes. *MNRAS*, **179**, 433

Camenzind, M. (1991). Magnetohydrodynamics of black holes and the origin of jets. *Annals of the New York Academy of Sciences*, **647**, 610

Dey, A., van Breugel, W., Vacca, W. D., & Antonucci, R. (1997). Triggered star formation in a massive galaxy at $Z = 3.8$: 4C 41.17. *ApJ*, **490**, 698

Di Matteo, T., Springel, V., & Hernquist, L. (2005). Energy input from quasars regulates the growth and activity of black holes and their host galaxies. *Nature*, **433**, 604

Förster Schreiber, N. M., et al. (2009). The SINS survey: SINFONI integral field spectroscopy of $z \sim 2$ star-forming galaxies. *ApJ*, **706**, 1364

Gaibler, V., Camenzind, M., & Krause, M. (2008). Large-scale propagation of very light jets in galaxy clusters, in *Extragalactic Jets: Theory and Observation from Radio to Gamma Ray*, ASP Conference Series (San Francisco), **386**, 32

Gaibler, V., Krause, M., & Camenzind, M. (2009). Very light magnetized jets on large scales – I. Evolution and magnetic fields. *MNRAS*, **400**, 1785

Häring, N., & Rix, H.-W. (2004). *ApJ Letters*, **604**, 89

Hawley, J. F., Beckwith, K., & Krolik, J. H. (2007). General relativistic MHD simulations of black hole accretion disks and jets. *Ap&SS*, **311**, 117

Heath, D., Krause, M., & Alexander, P. (2007). Chemical enrichment of the intracluster medium by FR II radio sources. *MNRAS*, **374**, 787

Krause, M. (2003). Very light jets. I. Axisymmetric parameter study and analytic approximation. *A&A*, **398**, 113

Krause, M. (2005a). Very light jets II: Bipolar large scale simulations in King atmospheres. *A&A*, **431**, 45

Krause, M. (2005b). Galactic wind shells and high redshift radio galaxies: On the nature of associated absorbers. *A&A*, **436**, 845

Krause, M., & Alexander, P. 2007. Simulations of multiphase turbulence in jet cocoons. *MNRAS*, **376**, 465

Krause, M. G. H. (2008). Jets and multi-phase turbulence. *Memorie della Società Astronomica Italiana*, **79**, 1162

McCarthy, P. J. (1993). High redshift radio galaxies. *A&A Review*, **31**, 639

McNamara, B. R., Kazemzadeh, F., Rafferty, D. A., et al. (2009). An energetic AGN outburst powered by a rapidly spinning supermassive black hole or an accreting ultramassive black hole. *ApJ*, **698**, 594

Miley, G., & De Breuck, C. (2008). Distant radio galaxies and their environments. *A&A Review*, **15**, 67

Müller, A., & Camenzind, M. (2004). Relativistic emission lines from accreting black holes: The effect of disk truncation on line profiles. *A&A*, **413**, 861

Nesvadba, N. P. H., Lehnert, M. D., De Breuck, C., Gilbert, A. M., van Breugel, W. (2008). Evidence for powerful AGN winds at high redshift: dynamics of galactic outflows in radio galaxies during the "Quasar Era". *A&A*, **491**, 407

Nulsen, P. E. J., Hambrick, D. C., McNamara, B. R., *et al.* (2005). The powerful outburst in Hercules A. *ApJ Letters*, **625**, L9

Tortora, C., Antonuccio-Delogu, V., Kaviraj, S., *et al.* (2009). AGN jet-induced feedback in galaxies – II. Galaxy colours from a multicloud simulation. *MNRAS*, **630**

Young, A. J., Lee, J. C., Fabian, A. C., *et al.* (2005). A Chandra HETGS spectral study of the iron K bandpass in MCG -6-30-15: A narrow view of the broad iron line. *ApJ*, **631**, 733

Ziegler, U., & Yorke, H. W. (1997). A nested grid refinement technique for magnetohydro-dynamical flows. *Computer Physics Communications*, **101**, 54

17

Cusp–core dichotomy of elliptical galaxies: the role of thermal evaporation

Carlo Nipoti

17.1 Introduction

There are two families of luminous elliptical galaxies: cusp galaxies and core galaxies. Cusp galaxies have steep power-law surface-brightness profiles down to the centre (hence the name 'power-law' galaxies, often used as a synonym for cusp galaxies), corresponding to intrinsic stellar density profiles with inner logarithmic slope $\gamma > 0.5$; core galaxies have surface-brightness profiles with a flat central core, corresponding to $\gamma < 0.3$ (Faber *et al.* 1997; Lauer *et al.* 2007). Cusp galaxies are relatively faint in optical, rotate rapidly, have disky isophotes, host radio-quiet active galactic nuclei (AGN) and do not contain large amounts of X-ray-emitting gas; core galaxies are brighter in the optical, rotate slowly, have boxy isophotes, radio-loud AGN and diffuse X-ray emission (for a summary of these observational findings see Nipoti and Binney 2007; Kormendy *et al.* 2009, and references therein). The most popular explanation for the origin of such a dichotomy is that cusp galaxies are produced in dissipative, gas-rich ('wet') mergers, while core galaxies are produced in dissipationless, gas-poor ('dry') mergers (Faber *et al.* 1997), the cores being a consequence of core scouring by binary supermassive black holes (Begelman *et al.* 1980). The actual role of galaxy merging in the formation of elliptical galaxies is still a matter of debate (e.g. Naab and Ostriker 2009). What is reasonably beyond doubt is that cores must be produced by dissipationless processes, while cusps are a signature of dissipation (Faber *et al.* 1997; Kormendy *et al.* 2009, and references therein). The dissipationless mechanism that forms the cores may not necessarily be scouring by binary black holes: even simple collisionless collapse may work (Nipoti *et al.* 2006). Dissipative processes must be invoked to explain the observed dichotomy because cusp galaxies are systematically less massive than core galaxies, and no purely stellar-dynamical mechanism can introduce a

AGN Feedback in Galaxy Formation, eds. V. Antonuccio-Delogu and J. Silk. Published by Cambridge University Press. © Cambridge University Press 2011.

characteristic mass scale. An interesting question is, therefore, why the dissipative process responsible for the formation of cusps works in lower-mass ellipticals, but does not work in the most massive ellipticals. Nipoti and Binney (2007) argued that efficient thermal evaporation of cold gas by the hot interstellar medium of the most massive ellipticals can be at the origin of the cusp–core dichotomy. Here I briefly describe the basic principles of this scenario and discuss some implications for the properties of AGN in elliptical galaxies.

17.2 The formation of cusps and cores in elliptical galaxies

The stellar-dynamical process that forms the cores is expected to operate at all mass scales: it is therefore natural to start from the hypothesis that all ellipticals at some stage in their evolution have central cores. Such cores can be later refilled by central bursts of star formation, but a necessary condition for a central starburst to happen is the availability of cold gas in the inner galactic regions. Accretion of cold gas into galaxies, from minor mergers or cosmic infall, is believed to be common even at late times. If nothing prevents accreted cold gas reaching the galaxy centre, central starbursts are likely and the core can be easily refilled by a cusp. However, the journey of an accreted cold-gas cloud from the outskirts down to the centre of an elliptical galaxy might be not so safe. Massive elliptical galaxies are embedded in haloes of hot (virial temperature) gas, and the interaction of cold ($T \lesssim 10^4$ K) gas clouds with such a hot interstellar medium can disrupt the clouds via a combination of ablation and evaporation by thermal conduction. The motion of a cold gas cloud through a hot plasma is a complex dynamical process, involving heat conduction, radiative cooling, ram-pressure drag and ablation through the Kelvin–Helmholtz instability. Less-massive clouds are more vulnerable, and it is possible to estimate with relatively simple analytic models a lower limit to the minimum mass a cloud must have to survive evaporation, depending on the temperature and density distribution of the hot interstellar medium (Nipoti and Binney 2007). The total mass of cold gas available for central star formation is determined by the mass spectrum of accreted gas clouds and by the minimum mass for survival against evaporation. The aggregate mass of new stars formed as a consequence of cold infall is estimated to be proportionally larger in lower-mass (hot-gas poor) ellipticals than in higher-mass (hot-gas rich) ellipticals (Nipoti and Binney 2007). Thus cores are likely to be refilled in the former, but not in the latter. This is consistent with the fact that all galaxies with high X-ray emission from hot gas are core galaxies, while ellipticals with lower X-ray emission include both core and cusp galaxies (Pellegrini 2005; Ellis and O'Sullivan 2006).

If cusps are formed by late starbursts that refill a pre-existing core, the central stellar population in cusp galaxies must be relatively young. This prediction is

C. Nipoti

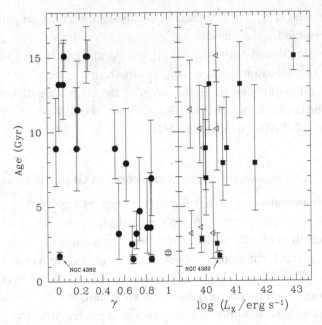

Figure 17.1 The age of the central stellar population versus the central slope of the intrinsic stellar density profile γ (left-hand panel; adapted from Nipoti and Binney 2007) and versus the soft X-ray luminosity L_X (right-hand panel) for early-type galaxies studied by McDermid *et al.* (2006). Central ages (with error bars) are from McDermid *et al.* (2006), the values of γ are from Lauer *et al.* (2007) and the X-ray luminosities are from Pellegrini (2005) and Ellis and O'Sullivan (2006). Measures of both L_X and γ are not available for all the galaxies, so two slightly different sub-samples are represented in the left and right panels. In the right-hand panel solid squares represent detections and empty triangles upper limits in X-rays. The empty symbol in the left-hand panel represents NGC 4459, which is not in the Lauer *et al.* (2007) sample, but is classified as a cusp galaxy by Kormendy *et al.* (2009): in the plot $\gamma = 1$ is assumed arbitrarily just to indicate that it is a cusp galaxy.

nicely verified in the sample of early-type galaxies for which McDermid *et al.* (2006) estimated the age of the central stellar populations: in the plane of central age versus central logarithmic slope γ (Figure 17.1, left-hand panel) early-type galaxies from this sample are found to be neatly segregated. Core galaxies have median central age 13.2 Gyr, while cusp galaxies have median central age 3.6 Gyr (Nipoti and Binney 2007). Given the known anti-correlation between γ and the soft X-ray luminosity L_X (galaxies with high L_X have $\gamma \lesssim 0.3$; Pellegrini 2005; Ellis and O'Sullivan 2006), it is interesting to see how galaxies from the same sample are distributed in the plane of central age versus L_X (Figure 17.1, right-hand panel): consistent with the proposed scenario, there are no points in the bottom-right area of the diagram. In other words, among the galaxies of this sample, all those with young ($\lesssim$5 Gyr) central stellar populations have X-ray luminosity lower than $\sim 3 \times 10^{40}$ erg s^{-1}.

The only outlier of the bimodal distribution in the age–slope plane is the lenticular galaxy NGC 4382, which is classified as a core galaxy, but has a young central stellar population (Figure 17.1, left-hand panel). It must be noted that NGC 4382 is also quite peculiar in other respects: its very unusual morphology and surface-brightness profile suggest interpreting it as an unrelaxed recent merger (Kormendy *et al.* 2009). Moreover, the diffuse X-ray luminosity of NGC 4382 is relatively low (Sivakoff *et al.* 2003; Pellegrini 2005), so the occurrence of a central starburst is not surprising. One might speculate that in NGC 4382 we are witnessing the first stages of the core-refill process.

17.3 Implications for active galactic nuclei in elliptical galaxies

Thermal evaporation may also have a role in determining the mode of accretion of central supermassive black holes in early-type galaxies. It is widely accepted that there are two main modes of black-hole accretion and feedback, usually referred to as 'cold mode' (or QSO mode) and 'hot mode' (or radio mode; e.g. Binney 2005; Hardcastle *et al.* 2007). In the cold mode the black hole feeds from cold gas, close to the Eddington rate, and grows significantly in mass, with most of the energy released going into photons (optical, UV, X-ray). In the hot mode the black hole feeds from hot gas, at a rate much below Eddington's, and does not grow significantly in mass, with most of the energy released being mechanical and generating significant radio emission. Bright QSOs and central radio sources in galaxy clusters are prototypes of the cold and hot modes, respectively, but the basic principles of this classification of the accretion modes apply to AGN in general. In the proposed picture cold gas can be available for accretion onto the central black hole only in cusp galaxies. Thus, all core galaxies must be hot-mode accretors, while cusp galaxies can be cold-mode accretors, when there has been a recent episode of cold gas infall into the galaxy. This is consistent with the findings that the optical nuclear emission (in units of the Eddington luminosity of the central black hole) is typically higher by two orders of magnitude in cusp galaxies than in core galaxies, that the nuclei of core galaxies are radio-loud, while those of cusp galaxies are radio-quiet (Capetti and Balmaverde 2006; Balmaverde and Capetti 2006), and that only core galaxies appear able to produce powerful extended radio emission (de Ruiter *et al.* 2005). This model fits a more general scenario in which radio-loudness is controlled more by the accretion mode than by black-hole spin, and hot-mode AGN, similar to the micro-quasars observed in our galaxy, alternate short bursts of radio-loudness with longer periods of radio-quiescence during their lifetime (Nipoti and Binney 2005; Nipoti *et al.* 2005).

17.4 Conclusions

Since the discovery of the cusp–core dichotomy of elliptical galaxies, it has been proposed that cores are formed by dissipationless processes, such as binary black-hole scouring, while cusps are formed by dissipative processes, such as merger-driven central starbursts (Faber *et al.* 1997). This proposal is consistent with several observed properties of core and cusp galaxies, but does not explain, per se, why the most massive ellipticals are cored, while less massive ellipticals are cusped. A possible explanation is that all ellipticals originally have central cores: in less-massive (hot-gas poor) ellipticals the cores are refilled by central starbursts following cold gas infall, while in more massive (hot-gas rich) ellipticals the cores are preserved because the hot interstellar medium ablates and evaporates most of the infalling cold gas (Nipoti and Binney 2007). In this scenario, black holes in core galaxies always accrete from hot gas, while black holes in cusp galaxies can accrete from cold gas, consistent with the observed properties of AGN in elliptical galaxies. The importance of the hot gas in preserving the cores in the most massive systems is supported by the state-of-the-art study of the early-type galaxies in the Virgo cluster by Kormendy *et al.* (2009).

In the present paper attention has been focused on the role of thermal evaporation in determining the cusp–core dichotomy of elliptical galaxies, but elimination of accreted cold gas by the hot interstellar medium, via ablation and heat conduction, is likely to be a fundamental process for galaxy formation in general, as it may be at the origin of the truncation of the blue cloud and of the population of the red sequence in colour–magnitude diagrams of galaxies (Binney 2004; Nipoti and Binney 2004, 2007). In this picture, the role of black holes in quenching star formation is fundamental, but indirect: via AGN feedback they supply the hot gas with the energy necessary to evaporate the cold gas and thus quench star formation.

Acknowledgements

I am grateful to James Binney for his comments on the manuscript and to Marc Sarzi for useful discussions.

References

Balmaverde B. and Capetti A. (2006). *A&A*, **447**, 97
Begelman M. C., Blandford R. D. and Rees M. J. (1980). *Nature*, **287**, 307
Binney J. (2004). *MNRAS*, **347**, 1093
Binney J. (2005). *Phil. Trans. R. Soc. London, A*, **363**, 739
Capetti A. and Balmaverde B. (2006). *A&A*, **453**, 27
de Ruiter H. R., Parma P., Capetti A. *et al.* (2005). *A&A*, **439**, 487

Ellis S. C. and O'Sullivan E. (2006). *MNRAS*, **367**, 627

Faber S. *et al.* (1997). *AJ*, **114**, 1771

Hardcastle M. J., Evans, D. A. and Croston J. H. (2007). *MNRAS*, **376**, 1849

Kormendy J., Fisher D. B., Cornell M. E. and Bender R. (2009). *ApJS*, **182**, 216

Lauer T. *et al.* (2007). *ApJ*, **664**, 226

McDermid R. M. *et al.* (2006). *MNRAS*, **373**, 906

Naab T. and Ostriker J. P. (2009). *ApJ*, **690**, 1452

Nipoti C. and Binney J. (2004). *MNRAS*, **349**, 1509

Nipoti C. and Binney J. (2005). *MNRAS*, **361**, 428

Nipoti C. and Binney J. (2007). *MNRAS*, **382**, 1481

Nipoti C., Blundell K. M. and Binney J. (2005). *MNRAS*, **361**, 633

Nipoti C., Londrillo P. and Ciotti L. (2006). *MNRAS*, **370**, 681

Pellegrini S. (2005). *MNRAS*, **364**, 169

Sivakoff G. R., Sarazin C. L. and Irwin J. A. (2003). *ApJ*, **599**, 218

Index

3C 236, 64
4C31.04, 169

Abell
 radius, 9
 richness, 4
absorption lines, 79, 98, 157
 MGII, 98, 101
 QSO, 100, 101
accretion
 ADAF, 90, 91
active galaxies, xiii
active galactic nuclei (AGN), 111
 activity, xiii, xiv, 11, 12, 17, 19, 21, 24, 25, 28, 31,
   32, 44, 67, 82, 83, 90, 128, 158, 159
 feedback, xiii, xiv, 30, 36, 55, 56, 83, 86, 89, 90, 91,
   92, 93, 94, 95, 98, 111, 112, 151, 152, 153, 154,
   171, 173, 175, 176, 177, 178, 179, 180, 181, 198
 obscured, 22, 24, 25, 30, 31, 65, 77
 Type I, 139
 Type II, 139

black hole, 82, 90, 91, 92, 96, 184, 191, 192
BPT diagram, 84
broad absorption line system (BAL), 99

Cen A, 64, 69
cluster of galaxies, 4, 7, 8, 175, 176, 178, 179, 180,
 198
 color–magnitude, 30, 33, 56, 111, 137, 144
 cooling flow, 11, 44, 82
 intracluster medium, 14, 15, 17, 29, 67, 82, 89, 181
 luminosity function, 83, 158, 161
Compton
 inverse, 76, 85
 self-synchrotron, 76, 78
cooling function, 113, 129

Doppler, 77
dust, 22, 24, 44, 67, 70, 190
 extinction, 142
 reddening, 67, 72

early-type galaxies, 3, 4, 6, 10, 63, 64, 128, 129, 196,
 197, 198
Eddington
 luminosity, 54, 67, 72, 197
 rate, 90, 96, 177, 192, 197
electron densities, 18, 130, 131
emission line, 67, 71, 135, 139, 157, 165, 188,
 189
 broad lines, 32, 65, 68, 76, 157
extinction, 142

Fanaroff–Riley (FR), 93, 96
feedback
 AGN, 3, 16, 17, 41, 42, 43, 44, 45, 68, 72, 73, 146,
   147, 175, 176
 negative, xiv, 3, 45, 93, 111, 131, 152, 175,
   176
 positive, xiv, 111, 112, 132, 136, 152
filling factor, 13, 161, 162, 173, 186

galaxies
 downsizing, xiv, 42, 45, 83, 88, 93, 149, 154
 formation, 41, 43, 44, 45, 63, 83, 88, 89, 198
 hosts, xiv, 25, 32, 33, 34, 36, 63, 64, 67, 71, 77, 83,
   95, 98, 99, 101, 102, 104, 105, 107, 111, 134,
   139, 154, 158, 183, 184
 star formation, xiii, 3, 4, 9, 21, 29, 30, 31, 32, 35,
   36, 43, 44, 45, 55, 56, 70, 71, 73, 93, 131, 132,
   147, 175, 176, 178, 181, 183, 191, 195, 198
 starburst, xiv, 24, 25, 26, 43, 70, 105, 172, 195, 198

IC 5063, 65, 67, 68, 72
IERS B0108+388, 77, 78
IERS B0500+019, 77
IERS B0710+439, 77
IERS B0941-080, 77
IERS B1345+125, 77
IERS B1404+286, 77
IERS B2120+048, 77
IERS B2352+495, 77
intergalactic medium, 63
ionization, 67, 90

200

Jeans mass, 116
jets, 29, 63, 64, 72, 75, 76, 83, 85, 86, 87, 88, 90, 91,
    96, 98, 111, 115, 119, 124, 132, 133, 159, 171,
    183, 184, 185, 186, 187, 190
  collimation, 165, 166
  jet-induced star formation, xiii, 65, 67, 69, 70, 72,
    112, 130, 171, 172
  relativistic, 64, 77, 159, 165, 168, 171, 184

LINER, 139
luminosity function, 82, 158
Lyman alpha system, 183
Lyman limit, 100

Minkowski object, 70, 73

narrow absorption line system (NAL), 98, 99, 100,
    101, 102, 103, 106
narrow line region, 76
NGC 4382, 197
NGC 4459, 196
NGC 541, 70
nuclear activity, 43, 63, 64, 65
nuclear disc, 54, 83, 89, 90, 91, 166, 168, 169, 184,
    190

outflows, xiv, 29, 68, 105, 158, 159, 160, 165, 171,
    172, 190, 191

PKS 1549-79, 72
PKS 2250-41, 73
PKS B0941-080, 77
power spectra, 49

QSO
  absorber, 98, 99, 100, 102, 103, 104, 105, 157
  feedback, 98, 99, 128, 147
  quasar, 22, 24, 25, 29, 43, 44, 45, 98, 99, 100, 101,
      103, 104, 105, 116, 197
    radio loud, 63, 105
    radio quiet, 105

radiation
  synchrotron, 76, 77, 78, 80, 85, 86
  thermal, 77, 78, 79, 80

Seyfert galaxies, 139, 157
Sloan Digital Sky Survey, 111
star formation rate, 29, 56, 57, 69, 70, 89, 92, 93, 94,
    96, 112, 177, 191

ULIRG, 22, 23, 34

X-ray, 19, 20, 78, 79, 80, 151, 157, 185, 195,
    197
  line emission, 78, 194, 195
  soft, 196

Printed in the United States
by Baker & Taylor Publisher Services